Project Partnering

Principle and Practice

Other books on similar subjects by the author

Management Applied to Architectural Practice
(with John Brunton and Eric Boobyer), Godwin Press, 1964

Managing Metric Change – A Management Action Plan
(with J. V. Connolly), Kogan Page, 1971

Training for Change
(with J. V. Connolly), Wellens Publishing, 1972

Managing Construction Conflict
Longmans Scientific & Technical, 1988

Total Quality in Construction Projects
Thomas Telford, 1993

. . . and manuals by Polycon under his editorial control

Quality Co-ordinator's Guide
(with the College of Estate Management), 1987

2nd Party Project Auditing, 1991

Project Partnering Manual, 1994

Project Partnering

Principle and Practice

Ron Baden Hellard

Partnering is the master key element for project quality. It opens the doors to commitment and communication between the parties and provides the mastic which turns disparate groups with varying aims into a coherent team with common objectives.

Published by Thomas Telford Publications, Thomas Telford Services Ltd,
1 Heron Quay, London E14 4JD.

First published in 1995

Distributors for Thomas Telford are
USA: American Society of Civil Engineers, Publications Sales Department, 345 East 47th Street, New York, NY 10017-2398
Japan: Maruzen Co. Ltd, Book Department, 3–10 Nihonbashi 2-chome, Chuo-ku, Tokyo 103
Australia: DA Books and Journals, 648 Whitehorse Road, Mitcham 3132, Victoria

A catalogue record for this book is available from the British Library

ISBN: 0 7277 2043 0

Produced by Gray Publishing, Tunbridge Wells, Kent
Printed in Great Britain by Redwood Books, Trowbridge, Wilts

Contents

Acknowledgements

In the three years since *Total Quality in Construction Projects* was written (to which this book is virtually a companion, if not a successor volume), the number of firms certified against the system set out in ISO 9000 has more than doubled. The first European and UK awards for quality management have been made and there has been substantial interest and growth in the application of the concept and its techniques in manufacturing, process and service industries and even in public service sectors.

I am grateful, therefore, to Thomas Telford for their agreement to frequently refer to, and sometimes reproduce parts of the earlier work, so that this book dealing with partnering principles and practices, may stand on its own for new readers. For those who have read the earlier work I hope this repetition will not annoy but merely reinforce – and improve – their understanding of the message.

I have, as before, drawn on work over the years with clients and with my colleagues in both executive construction and in consultancy. Without their work – and their frequent forbearance – this book would not have been possible.

I am grateful to Doug Pruitt, President of Sundt Corporation, Tucson and the President of Associated General Contractors of America, Washington DC, who also provided the information on US practices and for permission to include the conditions for The Marvin M. Black Award for Project Partnering.

I am also indebted to Roger Quick for information on Australian practices, and to Denis Wilson of the Federation of Master Builders of Australia for details of the Australian Award; to HMSO and Sir Michael Latham, for permission to reproduce parts of *Trust & Money* and *Constructing the Team*; to the British Quality Foundation for material relating to the UK Award, and to the Department of Employment for the 'Investors in People' diagram on pages 24–5. I also owe thanks to Diana Porter whose word processing skills have continued to make the gestation of this work easier and more enjoyable as, collectively, our own performance on its production has improved. It has indeed been *Partnering in Practice*!

List of figures and tables

About the author

Ron Baden Hellard has been involved in the construction industry at top level for over 40 years. He was the first Chairman of the RIBA Management Committee (which developed the *Management Handbook*), and President of the London Society of Architects in the 1960s. He has extensive experience in management consultancy in the UK and overseas and has advised many organizations on the management of change.

In the 1970s, Ron Baden Hellard developed training packages for the management of change and for construction professionals, contractors, manufacturers, government and service agencies. Since 1980, he has been involved in developing alternative techniques for the resolution of disputes, acting as an arbitrator, adjudicator and conciliator through Polycon AIMS Ltd.

He has provided total quality management consultancy to professionals, contractors and subcontractors in the building industry on over 100 assignments and has presented public seminars in Denmark, Norway, Holland, Hungary and Portugal as well as in the UK.

Ron Baden Hellard is Chairman of the Register of Consultants of the British Quality Foundation and on its Building Industry Steering Group.

Foreword

It is a great pleasure to review Ron Baden Hellard's new work on project partnering. He is tackling an approach to construction which has been regarded with doubt, scepticism or even outright suspicion by some clients in the past. But attitudes are changing rapidly. Clients now see themselves increasingly as at the core of the construction process, not as passive observers who are there to receive what the industry supplies for them and then pay for it.

Partnering is the affirmation by all in the construction process that they want to work together. It should not be a cosy arrangement.

A properly structured partnering agreement may very well have originated from a selective competition by the client to find one or more parties. If the client is in the public sector, or otherwise covered by European Union rules, it will almost certainly have included an initial process of that kind. There is no reason why partnering should be limited to specific kinds of construction, or large projects. It should also ripple down the entire process, in that a main contractor involved in a partnering agreement with a client should encourage similar relationships with the sub-contractors who will actually do most of the site work. Indeed, the partnering client may well insist upon such an arrangement, as part of the agreement, and thereby ensure that all involved in the scheme are fully meshed into the process on a basis of teamwork.

A site which is gripped by conflict, or adversarial attitudes, is of no benefit to the client. But clients have responsibilities as well as rights. Much attention is given in 'constructing the team' to ensuring that clients have the information and guidance necessary to launch a project efficiently. But, once the process is under way progressive clients will wish to ensure that all who are involved – consultants, contractor and sub-contractors – understand their objectives, and are working to the same end.

Partnering cannot mean the client detaching from the process and leaving it all to others. Nor should the client be uninterested in ensuring that all are paid on time. Prompt payment begins with the client, and should be followed by the main contractor in accordance with the contract. My report suggests remedies for failure to pay promptly, which Ron Baden Hellard also addresses in this book.

But I hope that those remedies will be rarely used. Teamwork involves solving problems together, not calling in referees all the time to solve them for the participants. Teamwork also means quality – and total quality should infuse the entire process.

If the objective of TQM is to be achieved, it needs everyone working together. There is too much defensive talk about defects being inevitable, with the excuse that projects are 'bespoke' and each job is different. Does any other industry think on that basis? Why should 'right first time' not be applicable in construction? And if we do not believe in our own abilities to control production in the interests of quality why should everyone else? Clients have only so much cash available to spend on their projects. If too large a proportion is frittered away in delays due to rectification of defects, less will be available for commissioning further work. Quality makes financial sense, as well as being a duty owed to the client by all in the process. Conflict is expensive, and nobody gains by it.

I am delighted that Ron Baden Hellard has once again led discussion on a vital issue for the industry. I hope his practical advice will be widely studied and followed. It is a vital part of 'constructing the team' and that clients begin to take the lead to set the process in motion.

Sir Michael Latham, DL, MA
Chairman of the Construction Industry Board
London
February 1995

Acronyms and definitions

Acronyms

ADOT	Arizona Department of Transport
ADR	Alternative Dispute Resolution
AGCA	Associated General Contractors of America
BAA	British Airports Authority
BPF	British Property Federation
BQF	British Quality Foundation
CCF	Construction Clients Forum
CDM	Construction Design Management
CED	Chief Executive Officer
CIB	Construction Industry Board
D&B	Design & Build
DoE	Department of the Environment
DTI	Department of Trade and Industry
EDG	Employment Department Group
EFQM	European Foundation for Quality Management
FMBA	Federation of Master Builders of Australia
ICE	Institution of Civil Engineers
ISO	International Standards Organization
JCT	Joint Contracts Tribunal
NACCB	National Accreditation Council for Certification Bodies
RIBA	Royal Institute of British Architects
TQM	Total Quality Management

Definitions

Definitions related to quality and its management

Assessment register (DTI QA register) Maintained by the Department of Trade and Industry of firms which have been audited and assessed by an independent certification body as being of assessed capability.

Certification There are three types of certification.

- First-party certification – that is the certificate of warranty given by the seller (the first party to the contract) to the buyer.
- Second-party certification – a certificate after assessment of the seller's system by the buyer, or his or her representative. For construction this would generally be project specific.

- Third-party certification – that given by an agency not related to any specific contract. There are three types of third-party certification:

(*a*) *certification of a quality management system.* This applies to a firm involved in providing a product or service. The firm is then authorized to use the NACCB symbol in its literature

(*b*) *product-conformity certification.* In this case, in addition to certification under (*a*), the product will have been assessed and tested to ensure that it conforms to the kitemark, the BSI safetymark, and Certification Authority for Reinforcing Steel (CARES) schemes

(*c*) *product approval.* In this case, in addition to certification as in (*a*) and (*b*), the product is assessed for performance in use (e.g. British Board of Agrement Certificate).

Certifying body (third party) An independent certification body, which will itself have been assessed as a firm capable of reliably working to specification and having a management control system that will comply with ISO 9000. In the UK a recognized certifying firm will be accredited by the National Accreditation Council for Certification Bodies (NACCB).

Defect The non-fulfilment of specific requirements. Non-conformity may apply to one or more quality characteristics of a service of elements of a quality system.

Product or service This may be:

(*a*) the result of activities or processes (a tangible product; an intangible product, such as a service, a computer program, a design, directions for use)

(*b*) an activity or process (such as the provision of a service or the execution of a production process).

A building contract embraces both and a partnering charter may cover both.

Quality The totality of features and characteristics of a product or service that bear on its ability to satisfy stated or implied needs.

Quality assurance All activities concerned with the attainment of quality – a process designed to increase confidence in a product's or service's ability to achieve the stated objectives.

Quality audit Examination to determine whether quality activities comply with planned arrangements and whether these are implemented effectively to achieve objectives.

Quality control The operational techniques and activities which together sustain the product, service or quality specified.

Quality loop; quality spiral Conceptual model of interacting activities that influence the quality of a product or service in the various stages ranging from the identification of needs to the assessment of whether these needs have been satisfied.

Quality management That aspect of the overall management function that determines and implements the quality policy – on a building project synonymous with project management.

Quality management system The total management system of organizational structure and responsibilities, activities, resources and events which together provide on a building project procedures and methods of implementation to ensure the project organization meets the quality requirements.

Quality plan A document setting out the specific methodology, practices, resources and sequence of activities relevant to the particular project.

Traceability The ability to trace the application or location of an item or activity by means of recorded identification. Traceability may be applied to a product or service, the calibration of measuring equipment or calculations and data relevant to a product or service.

See also BS 4778 Glossary of terms used in quality assurance (including reliability and maintainability terms) (from which some of these terms have been taken).

Definitions used in relation to construction-related quality and partnering

Quality For construction, the needs must be defined by the client. The inclusion of services is pertinent to construction, where both designers and contractors supply services as well as the product (i.e. the completed work). The quality of these services is vital, not only in meeting the client's physical requirements, but also in completing to time and budget (function, aesthetic, cost and time – FACT) – quality requirements for any product.

State-of-the-art audit An initial audit done before a quality management programme is instigated (often second party). It should be concerned as much with the 'people culture' as it is with the existing paperwork, operating procedures, and the product or services' technical performance.

Building process The overall series of linked activities beginning with the identification of need and conception of the project and ending with the occupation and initial use of the construction after its commissioning.

Project The word used to indicate the overall activity being undertaken as opposed to the use of the term 'contract'.

Contract The legal document and framework under which the various parties carry out their work on the project.

Parties included in a building project and partnering process

Adjudicator A neutral person appointed by the parties to give, when requested by any party to the contract entered into related to a specific project, an interim award that is binding on the parties until the end of the contract when it may, if required by either party, be reviewed.

Client representative The person appointed by the client either from within his or her own organization or specifically appointed to be the single point of communication between the client and all parties to the contract, particularly for contractual instructions. The client representative would therefore normally be named for this purpose in all contracts subsequently entered into by the client.

Project manager A project manager may be appointed in addition to a client representative. The project manager will need to have the role defined and his or her authority, responsibilities and accountabilities indicated in the terms of appointment and then subsequently in all contracts entered into between the client and other parties to the contract. The use of a project manager does not eliminate the need for a nominated client representative but it will probably reduce the scale and scope of the client representative's duties.

Design professional An individual or firm providing a professional design service connected with the project in accordance with the code of conduct of one or other of the professional institutions. This covers architects, structural engineers, civil engineers, building surveyors, landscape architects, building services engineers and quantity surveyors. Any of these may be employed on a design contract, which may be individual for the technical area of design or may be a collective appointment to a firm offering a multidisciplinary service. The term distinguishes between the role as a professional and a similar role in which members of the disciplines are employed directly by a general or specialist contractor.

Facilitator A neutral person retained to assist the parties to establish the charter, educate the parties involved, organize and/or run the initial partnering workshop. The facilitator has no executive function within the contract, unless he or she subsequently acts as the adjudicator during the progress of the works.

General contractor A firm undertaking the principal activity of construction. It implies overall control of the construction works, the contract for which normally empowers the contractor to enter into subcontracts with 'domestic' sub-contractors on his or her own terms, or with specialist sub-contractors on terms which may be defined by the client's agents, whether or not the specialist sub-contractor is 'nominated' by the client.

Specialist contractor (or sub-contractor) A firm providing a specialist part of the building such as heating services, lifts or curtain wall envelope. As is the case for the general contractor, the specialist contractor may include technical design as well as manufacture and installation on site, sometimes including the necessary commissioning activities.

Craft contractor (or sub-contractor) A firm supplying particular craft skills such as decorative plasterwork, plastering, asphalting or classical stonework. Craft contractors are generally small firms and employed as 'domestic' sub-contractors (i.e. to a main contractor, and so with no direct relationship to the client).

1 Introduction

> Partnership is our great curse. We too readily assume that everything has two sides and that it is our duty to be on one or the other.
>
> *James Harvey Robinson*
> *The Mind in the Making*

> Whoever makes two blades of grass grow where only one grew before deserves better of man-kind than any speculative philosopher or metaphysical system builder.
>
> *Jonathan Swift, circa 1700*

The 'partnering' philosophy is the master key that will unlock the techniques and principles of TQM to provide customer satisfaction on construction projects.

The construction industry in the UK and in the English-speaking world generally has evolved in response to the specific needs of unique projects for individual customers. Design firms are generally separate from general building contractors, which then means that they are excluded from the specialist firms of sub-contractors for traditional trades – themselves directly descended from the journeymen craftsmen of the renaissance. Modern technology has developed specialist manufacturers for even larger components and elements of buildings which require their own specialist teams to erect or install, but again in separate firms.

These organizations, currently with less differentiation between their 'commercial' and 'professional' characteristics, develop their own operating skills which have in some cases included systems that comply with the requirements of ISO 9000 (and some, but relatively few, have had them certified). Of the 12,000 firms certified in the first 10 years fewer than 250 of the construction industry's approaching 100,000 firms were included. By 1994 the UK total of certified firms had reached 34,000, with less than 500 being firms involved in building – manufacturers excepted.

But, individual systems in each area of professionalism even if they are all based on ISO 9000 will not ensure a quality result on the client's project. This requires first a project-based system tailor-made for the location, specification, and client's specific requirements. Then it requires the development of common goals, unified objectives and teamwork to achieve them. A requirement that is essentially people orientated and one which with strong motivation, direction and management will develop a win/win project culture in which all stakeholders share and benefit. Management systems, whether based on the RIBA Plan of Work, the BPF system of contracts, or ISO 9000, can never achieve this alone. This is the task of the partnering philosophy and the challenge facing UK construction as it moves towards the 21st century.

In the 1990s the UK construction industry has suffered its most severe recession this century, with the biggest and many of the best contractors and design firms making losses in successive years, and many of the smaller firms (and not the worst) going out of business.

Litigation and arbitration over construction disputes have continued to grow and so perhaps, not surprisingly, the take-up of quality practices has been infinitesimal. Yet in the United States and Australia, where construction has also suffered severe depression, the technique and spirit of partnering has been successfully adopted.

That management systems like ISO 9000 can develop quickly there is no question, but human nature and habits that develop group cultures, essential to stable and comfortable living are much slower and more resistant to change. Change as an idea is good for others; the young, and for those who can be trained for it. But we, particularly those of generally more mature years and as a consequence in more senior and authoritative positions, are happy to accept the benefits as long as we do not have to alter our thinking, practices and lifestyle.

Enforced culture change invariably means a culture shock. Change is unsettling, creates fear of the consequences in those lacking confidence in themselves or in their ability to cope with what will be required of them in the future. Thus, change engenders resistance in people, and without their full and enthusiastic support the improved systems, methods, machines and materials will never achieve their full potential.

This is the challenge for industry as we move into the 21st century. However, the challenge facing the project-based construction industry, and in particular those engaged on the complex buildings required for modern living in the technological era of the next century, is more immediate.

Each project to be undertaken is unique. Unique in its time, location, physical requirements, standards required and, above all, in the human relationships needed to design, develop, construct, equip, commission and maintain. A process which begins with the client and his or her conception of requirements, but this is (rarely and properly) not a specification of needs.

Some clients who have a regular and continuing need to procure buildings have the economic incentive to develop and maintain expertise and professionalism in 'being a client'. However worldwide, most clients have this need but once in their lifetime and therefore may be considered, and frequently are, 'amateur' in the process. These gifted amateurs – and most are – must then be brought into the project team and educated into their role, which in essence is as its leader – the strategic decision-maker of what is to be built, when, where, and to what standards. Most important of all is the decision on 'who' is to be involved in the creation of the physical edifice to provide his or her project requirement. To embrace the Latham reports it is his money and he has to decide in whom he will put his trust and how he will construct the team to achieve his objectives. This is the particular role of 'partnering' and it is the main thrust of this book, as it examines the application of management philosophy and techniques to the creation of the client's project. Partnering, some may argue, is only the re-adoption of the approach of a more leisurely age. Architects and contractors took a gentlemanly approach to working together to please their clients. Frequently in a local, almost parochial environment, they gave their clients a simpler traditionally built solution to simpler requirements. Today, cynics say, life is faster, projects are more technically demanding, the sums involved are vast, competition is fierce and financial pressures to make a profit for shareholders are all demanding. So, catch as catch can and let the devil take the hindmost to outsmart the opposition is the name of the game – which must be won at all costs.

But the practical outcome of the successful application of quality management techniques allied to its philosophy and principles has shown, in industry across the world, that the quality company is the profitable company and that quality investment reduces costs. What is needed is the successful adaptation and adoption of these maxims to the project situation so that all the stakeholders in the project benefit, not least in the achievement of individual and team satisfaction, and so have fun in the process – what Deming called 'the joy of work'!

This is the aim of partnering. Partnering should include all the stakeholders which, particularly in relation to a construction project, includes society at large – or at any rate those thousands who will work or live in the project and the far greater number who will pass by or through it and be aware of the visual and even emotional impact it makes on their living environment. Partnering is the essential philosophical framework for the application of the principles and practices of TQM to the construction project. This book, therefore, examines these areas in turn and adds a series of case studies of successful usage. Construction, perhaps even more than most industrial activities, is a people business and the emphasis therefore should be on the human aspects rather than the technological, on the constructive elements of relationships rather than on the legal applications of doctrines for contractual performance. Contract forms have fought (and failed?) to keep up with the technical rate of change and speed required for construction. All in order to limit the cost of finance to highly geared borrowers, whether they be clients, whose funds are unproductive whilst tied up in incomplete projects, or contractors overstretched by unplanned delays arising from client's (or their architects' or engineers') late changes to requirements!

Construction will always, by its nature, be subjected to changes for the unforeseen – and unforeseeable and thus dispute or problem resolution concepts and procedures, must feature in all partnering agreements and arrangements. So, they are also considered in this book and within the competitive environment which is a present norm in the UK and elsewhere, at least in the English-speaking countries, but through which cooperation must also take place.

To consider partnering in isolation from the overall concepts of total quality management would be like studying a highly volatile fuel before knowing the nature of the engine and the vehicle it was intended to drive. So, Chapter 2 deals with the overall implications of TQM and the project context within which it must operate. (See also *Total Quality in Construction Projects*, Thomas Telford, Chapters 1–3.)

2 Total quality management in the construction industry

> The great revolution that takes place in the mental attitude of the two parties under scientific management is that both sides take their eyes off of the division of the surplus as the all-important matter, and together turn their attention toward increasing that size of the surplus until this surplus becomes so large that it is unnecessary to quarrel over how it is divided. They come to see that when they stop pulling against one another, and instead both turn and push shoulder to shoulder in the same direction, the size of the surplus created by their joint efforts is truly astounding. They both realize that when they substitute friendly cooperation and mutual helpfulness for antagonism and strife they are together able to make this surplus so enormously greater than it was in the past.
>
> *Frederick Winslow Taylor 1912*

This chapter sets the scene for the whole of the construction project partnering activity, it looks first at the evolution of quality through standards and the many contributors to management philosophy and then at the evolution of international awards for TQM. These areas are then related to the different groups involved in the design and implementation of a client's project requirements – professional firms, general contractors, sub-contractors and also to the all-important role of the client in developing the team through which his or her project's success will be achieved. It examines how TQM affects each firm in turn, and then how the project quality plan provides the project system which must operate within the partnering philosophy, the particular application of TQM that is needed to change the construction project's culture.

In today's conditions of international competition both the manufacturing industry and the government have recognized that quality and reputation for time and cost-performance are equally important. It was within this context and the growing volume of consumer protection and EU product liability legislation that the government introduced its national quality campaign with the publication in 1982 of Command

Paper 8621, Standards, Quality and International Competitiveness. This included the establishment of a National Accreditation Council for Certification Bodies who accredit organizations who in their turn certify that a firm maintains quality assurance procedures that comply with BS 5750 (and ISO 9000–9004 – CEN 29000–29004). This is known as third party certification and has been taken up by more than 30,000 organizations in the manufacturing industry and commerce who are now on the Register of Quality Assured Firms. It is the government's intention to use this list increasingly as a qualifying condition for government contracts. Already some large organizations are doing so.

The construction industry has, in many ways, different problems and management requirements from those concerned with the manufacturing industry. First, it is a project-based operation. Next, each project tends to be a unique, one-off operation and, finally, the operation of each of its component organizations – architect, engineer, constructor and manufacturing supplier – are totally interdependent of each other, but operate through a variety of contractual arrangements and procedures which are specific to each particular client's requirements.

As has been said, ISO 9000 is the UK national standard covering the complete spectrum of British industry, dealing with the provision of a management system to ensure and show that the service (or product) reliably provides a performance that matches its promises. Thus, these standards can make a positive contribution to improved performance.

The government's intention to generally and substantially improve standards in the face of increasing international competition was followed in 1992 by the foundation of the British Quality Foundation. This was established by the DTI providing financial assistance on a 50/50 basis with industry. A primary aim of this being to establish a UK award for excellence in quality. The first award in 1994 was based upon the European Foundation for Quality Management's award first made in 1992 (to Rank Xerox UK). This was in turn based upon the Malcolm Baldridge Award in the United States begun in 1987 and itself in a direct line of development from the Deming Award first made in Japan in 1952.

The British Quality Foundation also developed for the 1995 awards, a new category of award for organizations with fewer than 250 people and further variants appropriate to the public sectors of health and education. All of these awards recognize the need for quality systems dealing with process management. Of the total weighting in the award model, 14% is allocated to this factor, but the need for substantial and critical self-assessment before entry is emphasized.

The concept of TQM goes far beyond management systems related to the production process. It embraces the philosophy, principles, processes, practices and procedures of management to provide customer satisfaction in the goods and services provided by all parts of the organization.

The standards approach

ISO 9000 gives general guidance on the application of the standard for quality assurance systems and guidance on the selection and use of other standards in the series. ISO 9000 and 9004 are particularly relevant to guidance for internal management of quality systems. Although they were developed and written originally for the manufacturing industry and relate to the products of the manufacturing industry, much of the guidance given is sound advice for managers in any organization.

The introductory sections on the characteristics of quality systems, organizational goals and the determination of specific objectives for the organization, on the need for and definitions of management responsibilities and of the organization structure through which these will best operate, cover principles applicable to all organizations, and so move towards quality management, albeit within a production ambience.

Customer satisfaction framework

TQM develops the principle of customer satisfaction to embrace internal as well as external customers and, as illustrated by the award assessment model, includes concern for all the stakeholders of the undertaking:

- the direct customer for the service or product
- society at large
- the shareholders
- workers providing the output of the enterprise.

It does this through the philosophy that within the organization every group or individual provides services to another group or individual, which is its customer. Within the production chain each operator in an assembly process passes on the work to the next in line on the process, who is his or her customer. At its simplest, when a director dictates a letter to a secretary, the secretary is the customer, and when the work is returned to the 'boss', it is the 'boss' who is the customer. Each should attempt to provide the service as the customer would like to receive it. Thus, the accounts department services all the other departments, as do the purchasing and maintenance departments.

Construction implications

In the construction industry this concept requires that the traditional culture needs considerable change as the sub-contractor becomes the 'customer' of the main contractor; the main contractor becomes the customer of the quantity surveyor (for the bill of quantities); the quantity surveyor the customer of the architect/engineer designer, and the designer the customer of the project manager – or client!

These culture changes will have a profound effect upon the industry, and its forms of contract, and illustrate that the application of TQM may need further development to suit the project nature of construction organizations.

The nature of the organizations that have developed in the UK, and elsewhere, to meet and supply the construction needs of the industry's clients are so different from those to be found in the manufacturing industry and elsewhere that a fundamental reappraisal of the concepts, approach, framework, and certainly the wording of the ISO 9000 series is needed if the achievement of a similar quality objective is to result in these organizations.

Project quality

It is vital that this objective should be reached within each company or firm as without it the benefits of the right approach to project quality by contractors and sub-contractors will be severely limited and the building's owner will not obtain the quality building he or she expects at the correct cost, delivered by the date agreed.

In short, the owner's brief will not be met and the building team will have failed to deliver a quality product – a building fulfilling the functional, aesthetic, cost, and time requirements of the customer, that is one fully fit for the customer's purpose. There will be many reasons or excuses for these failures, too often paraded before a judge or arbitrator, illustrating that the (poor) 'quality costs' are only too frequently considered 'normal' thus underlining the need for quality management systems and the benefits they should produce.

The whole building industry is, and for many areas must always remain, labour intensive. The people carrying out the design functions may not be so responsive to the documentary controls and paperwork systems of quality assurance developed and applied successfully in manufacturing and some service industries. The TQM approach and the concept of partnering moves the management philosophy closer to construction, but a 'project-based model' may be needed to put the construction process on a level playing field with the rest of industry, commerce, and even non-profit organization's activities when it comes to awards for excellence and improved performance!

TQM and 'the model'

Definitions of TQM are still developing in many areas – each reflecting the particular concept and environment in which they are developed.

The following quote from the Latham Report is the working definition adopted by the Henderson Committee in 1992, which led to the formation of the British Quality Foundation.

> 'Total quality management is a way of managing an organization to ensure the satisfaction at every stage of the needs and expectations of both internal and external customers, that is shareholders, consumers of its goods and services, employees and the community in which it operates, by means of every job, every process being carried out right, first time and every time'.

This quote has been reduced by the British Quality Foundation to

> *'TQM is a customer centred philosophy of management concerned with every aspect of an organization's activity'.*

The draft addendum to ISO 8402 'Quality Vocabulary' defines TQM at 3.50 as:

> 'A way of managing an organization which aims at the continuous participation and cooperation of all its members in the improvement of:
>
> - the quality of its products and services
> - the quality of its activities
> - the quality of its goals
>
> to achieve customer satisfaction, long-term profitability of the organization and the benefit of its members, in accordance with the requirements of society'.

That there should not be unanimity in either definition or understanding is not surprising. There have been many contributors to the still evolving concept, described as the third industrial revolution. TQM is the framework against which progress into the 21st century is taking place; it is hoped that this will be an era when concern for people will balance those for production and profitability.

Table 1 lists those pundits or gurus who have all contributed significantly to these and identifies their principal contribution. Not all agree on all points made by others, and indeed there are some direct conflicts in application. There is, however, general agreement that TQM is a comprehensive approach to managing work based on:

- analytical evaluation of work processes
- development of a quality culture
- the empowerment of employees.

It is in this climate that the TQM model (Fig. 1) has been developed and defined as:

> 'A structured approach for managing a business to achieve the best results '.

Table 1. Contributors to the TQM philosophy

Deming	Fourteen points – seven deadly diseases
Juran	Trilogy – planning, control, improvement
Feigenbaum	Quality costs
Crosby	Zero defects – 14 steps
Ishikawa	Thought revolution – honesty
Taguchi	Variety reduction through quality System, parameters and tolerance design
Townsend	Rebellion against mindless rules
Augustine	Rules no substitute for sound judgement
Drucker	Management must make customers
Herzburg	k.i.t.a.
Maslow	Hierarchy of human needs
McGregor	X and Y theories
Baden Hellard	Start at the project's top Involve colleagues and customers Continually train and develop all
Cartwright	Nine key motivators
Standardization	DIN, CEN, ANSI, ISO (9000 Series)

For further detailed analysis see *Total Quality in Construction Projects* by the author, published by Thomas Telford, 1993

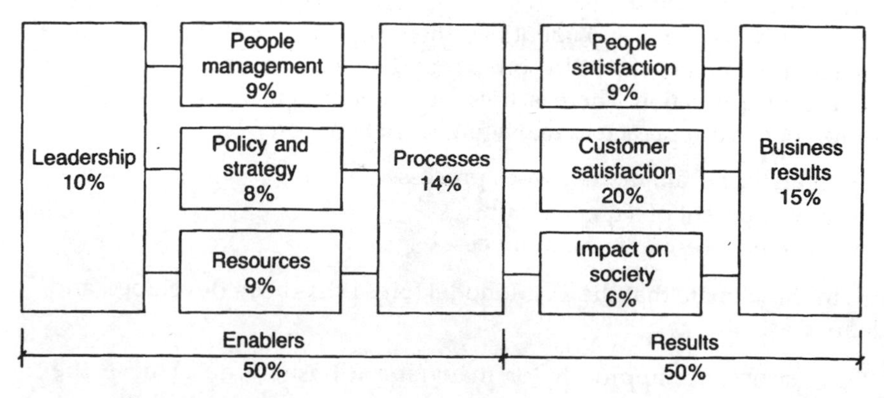

Figure 1. The TQM model

Construction people may understand the model and its implications better seen as in Fig. 2 , assuming it is of course built on the firm foundation of an established organization!

ISO 9000 has been described as the 'Doorstep to Quality' by Sir Derek Hornby who led Rank Xerox to win the first EFQM Award in 1992. British industry, he said, must get on the roof!

The needs of a construction design organization

Let us now examine some of the reasons why these needs are different from models suited for other organizations.

ISO 9001–9003 cover specifications for different types of manufacturing activity. ISO 9001 covers the most comprehensive range of activities and is the model for quality assurance in design/development, production, installation and servicing. ISO 9002 provides a model for firms which are concerned only with production and installation, the design having been carried out elsewhere, and 9003 provides a model for quality assurance where final inspection and testing only is involved. ISO 9004 deals with a service environment with multiple repeats of a customer interface, such as restaurants, hotels, insurance, and even hospitals and public services. Information, definitions and applications given in these specifications are not totally appropriate for professional organizations concerned only with building design. Nor are they appropriate for building firms carrying out projects under the very different forms of contract for construction

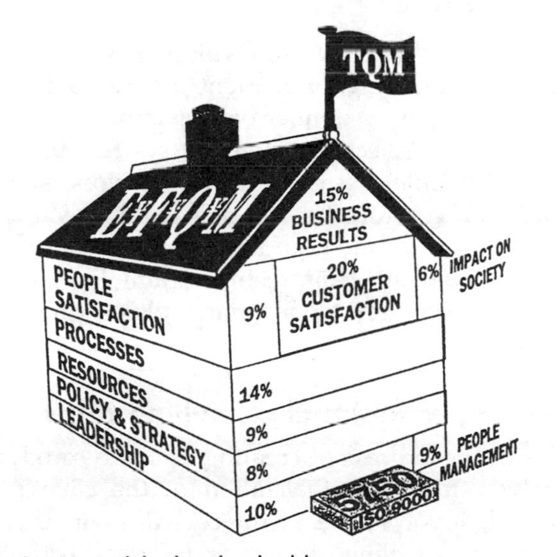

Figure 2. The UK/TQM model related to building construction

works. These areas are quite different from those involved in the simple buyer and seller relationship envisaged by the ISO series.

Different contractual context

In construction activities such organizations may be involved in production, construction, installation, etc. However some organizations may also be involved on a monitoring or advisory basis, with the contract for implementation of the work between separate third parties. Thus the context of a two-party contract as envisaged in the ISO 9000 series does not apply so well to either the professional consultant office, or the construction contractor, and sub-contractor.

Various forms of building and civil engineering contracts have evolved over the years to meet different situations. They are frequently poorly understood, badly administered, or their implications totally ignored until too late, but the substitution or overlay of a separate, or additional, documented layer of paperwork systems will only make the position worse, not better, and more expensive. The Latham Report *Constructing the Team*, published in July 1994, relates many of the acknowledged ills of the construction industry and quotes the author as follows:

> 'Improving management procedures at all levels is vital for the industry's performance, and needs to be part of a drive towards Total Quality Management (TQM). TQM is not solely about procedures. In his study of TQM, Ron Baden Hellard emphasizes certain aspects which lie at the heart of this Report. He stresses that the "philosophy of teamwork and cooperation, not confrontation and conflict, is long overdue" '.

The TQM approach is intended to continuously improve performance and encourage team building, implement project quality planning – an approach for post competitive tender partnership.

What is needed are project management systems providing a careful balance between 'people' systems and paperwork systems that will support (rather than control) the designers in their various firms if the outcome is to be a practical and economic achievement of the objective – a quality project. It is within this technical and legal context that the motivation to develop an appropriate philosophy has produced 'partnering'.

TQM systems in professional consulting firms

The first phase in the process of creating a quality product, the client's building, involves the conceptualization of the client's needs into a physical edifice. It involves the architect/civil engineer, and the other engineering design disciplines and their interface with the client and each other. Their systems and procedures must therefore cover the inception stage analysis of client needs. Akin to marketing in other industrial environments, this often requires also original research before the conceptualization can take place. When this has occurred, then, and only then, can forecasts of time and cost, and therefore the feasibility of the project, be made. The information for these forecasts is very limited, yet it is on this forecast that the client must make the fundamental strategic decision to proceed – or not. Quality systems in professional consultants' firms therefore require a careful balance between 'know-how' and experience, people systems and paperwork systems if the necessary control is to be achieved practically and economically.

Reasons for poor management in the professional context

It is sometimes said that professionals – designers, architects, engineers, surveyors, lawyers – are reluctant to embrace management control systems and apply management principles that are common elsewhere. This, it is suggested, is as a result of a self-opinionated, intellectual arrogance or aloofness. Regrettably, in my experience over 30 years objective analysis this is sometimes true, but there are many reasons why 'professional' activities have developed in the way they have.

Let us look at some of the principal reasons:

- Their contract specifically requires research, analysis, problem-solving and then design effort resulting from these activities when the contract requirements can only be stated in outline performance terms or where the standard of solution, advice or service cannot be established before the contract has been entered into.
- A professional service is one where the firm supplying the service is made up of individuals who have obtained a recognized qualification, have the necessary experience, have a requirement to comply with a code of conduct or a publicly recognized ethical code, such as applies to membership of professional institutions.
- Their professional training has developed an inbuilt desire for excellence – if not perfection – in the solution to the problem set, often to the exclusion of practical cost or time limits.
- Until very recent times most professional design offices have been made up of all small groups where the executive work is done by the most highly qualified member of the group with a small band of assistants under direct and constant supervision.
- The monitoring of the problem and its solution by another equally highly qualified professional would severely increase, if not double, the cost of the unique solution (and then might even require a third to arbitrate between two solutions!).

- Problem-solvers of unique problems develop an understanding of when they need to seek a 'second opinion' or help with a particularly difficult problem-area. Paperwork systems *could* be developed to monitor this kind of work but would be astronomical in cost. There would be no more guarantee they would be used or be better than relying on the first problem-solver recognizing his need for further advice!

  For these reasons paperwork monitoring systems in crucial problem-solving design areas will be little more than checklists for the problem-solver, and these will probably be those developed by him at an early stage of the work.
- By the conceptual nature of these design stages the 'brief' or statement of requirements is one that develops as an interactive process between the lead designer and the client – or building owner – who must become an essential component in the quality system. Indeed the project quality plan must belong to the client if it is to have any realistic meaning and chance of success. Thus identifying a further reason for the evolution of the present organizational framework for the industry.
- Most clients are 'amateurs' – hence the need for 'professional' consultants to solve the problems for them. Again, it is sometimes said that professionals get the clients they deserve (and vice versa), and, again, this is probably true if a client is allowed to be a bad client. A professional's first function, certainly as far as the quality plan is concerned, is to see that the client does what is necessary; this should be as a result of being made aware of what is expected of him or her and what is not, as well as what will and will not be undertaken by the construction team. It is this failure to manage the client that leads to most claims and disputes in building projects. This is not to say that the client does not have the right to a change of mind at any time and to get what is wanted; however the client must accept the implications and meet any extra contractual costs that may arise. Experience suggests that, if the client is made aware of these implications at the time the changes are requested, he or she might not go ahead anyway! The partnering concept aims to ensure that all of these interfaces and decisions are participative between the stakeholders who will benefit – or lose – by their implementation.

Requirements, resources and authority

The design and construction of a building or civil engineering project is, worldwide, one of the most complex and difficult undertakings. It requires management skills of a high order and is frequently undertaken by firms with little or no formal training in management.

To complete the construction fit for its purpose, correct first time, and to pre-determined quality standards which must therefore embrace the client's functional, aesthetic, cost limits and time requirements (the FACTs of the project brief) around which the project quality plan must be framed, requires not only planning, organizing, budgeting, controlling and careful adherence to an overall project management system, but also a high level of understanding of the industry's culture, of human motivation and behaviour within prescribed contract conditions.

It requires also recognition that quality management systems are never fully within the control of any one organization involved in the project's contracts. For, in the UK, as well as in most other countries in the world, other organizations and individuals can, and do, exercise authority and control over the project, but without accountability to the client, who is the building team's customer, for the overall performance of his or her requirements.

This control may affect the appearance, size and general layout of the project through town planning, fire and means of escape regulations – the combination of materials through building regulations – and the method of building through Health and Safety, CONDAM, traffic and noise control regulations. In many cases this control is exercised after the event, and, in some cases the nature of these authorities' particular requirements cannot be established beforehand. (See Fig. 3, taken from Chapter 2 of *Total Quality in Construction Projects* where the authority of outside agencies is dealt with in more depth.)

Principles of TQM

The principles of TQM developed within the context of a simple buyer and seller relationship – the first and second parties to the contract – need further development for the complex nexus of contractual relationships that must be organized to meet the unique requirements of a one-off project carried out on a unique site at a single point in time through a unique combination of corporate and human relationships.

Experience has shown that the technical failures of product or material are minute compared with the enormous dissatisfaction caused through failure to meet cost and time targets and human expectations, which are all part of the make-up of a quality project.

The client's role

Many of these failures stem from the failure to recognize the extent to which the building client should participate not only in the early stages of the inception of the project in fully exploring and analysing with the designer the full requirements, but also then, and subsequently, in

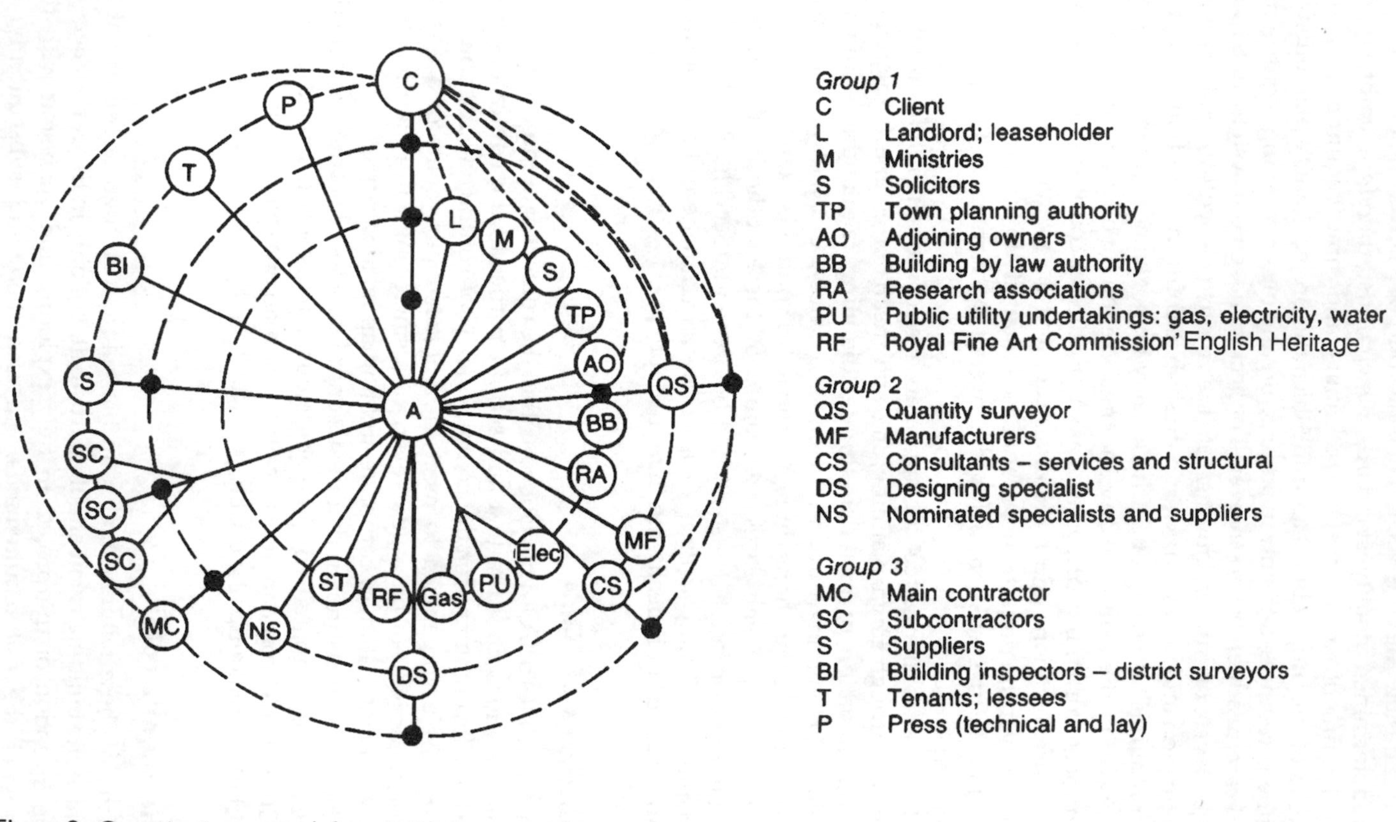

Figure 3. Organization network for a building project

making a full study of the ability, and suitability of the resources of those organizations through whom the client's requirements will be fulfilled. During the development of the Latham Report in 1993 and 1994, 'the client's role' was identified and client groups were encouraged to acknowledge its importance and to see their membership of the construction team in the light of TQM and partnering. It would appear a few 'regular' organizations of clients have done so. But still most building projects do not have 'regular' but rather 'one-off' clients and so the task remains to establish TQM at the earliest possible moment in the evolution of the construction project. For as Sir John Egan, originally responsible for developing the approach at Jaguar, and now Chairman of the British Airports Authority, has said: 'the earlier in the development of the project you pay the price of quality the greater will be the benefit'.

The client's role is, however, the most important one in establishing the team that is to carry out this project. Leadership of that team is vital, but only in the case of 'professional' or regular clients is the leadership likely to come from the client or client body. The issue of leadership must, however, remain the responsibility of the client and it is in the client's own self-interest to see that a project manager, or leader is appointed. The project manager should then be concerned with the creation of the right environment for partnering and facilitate effective team building on the client's behalf.

Every building client must recognize that in construction, perhaps even more so than with any service industry, the service received will depend in part on his or her own involvement in, and contact with, the supplier of the services. However much is delegated, and certainly the technical aspects of design and the bulk of the physical aspects of construction will be delegated, the client cannot delegate the ultimate responsibility for all key decision-making related to the project. The client must retain the decisions for what is built, where and when it is built, to what standard or level of performance, and within what cost and time limits, and finally and most important of all who is selected to implement the project.

In the context of construction, cost-effectiveness must be judged (unless the brief specifies otherwise) by the ultimate quality and costs-in-use of the building over its useful life-span – not by the cost of the planning, early design and organization of the project.

The earlier the cost of obtaining quality is built into the design and management process the greater will be its effect. Thus, the sensible client will seek a design team and design management resource that have established his or her own management systems and can illustrate them with hard evidence (and probably reputation). The client should then be prepared to invest in more – and not less – than the standard fee scales to carry out the fuller and more effective investigation into construction and construction resources that will produce a lower priced but better value building from a better motivated construction team, who will not then have to price unnecessary risk into the project.

Elimination of unnecessary risk is vital in a one-off project. Quality management and TQM systems aim to completely achieve this through effective second-party auditing. Such an audit or initial assessment must be project specific and will certainly pay off in the quality of the work and its cost, whether a contract is finally obtained by competitive or negotiated tender. Then, when the contract has been placed, the practices of partnering with the implementation team can begin. The client in the role of top manager of his or her building team, if not its leader, should begin it.

The project quality plan

The practices of partnering should begin through the project quality plan which will have its first implications on the work of the project manager and design team, upon whom the principal implications of TQM have already been explored in this chapter.

An essential requirement for successful quality in construction (and of ISO 9000) is the preparation of a plan for every project. The motivation, instruction and guidance given by quality management gurus and publications generated is concerned with creating within the firm a TQM system. It has also been concerned with motivating and developing individuals within the firm to realize and exercise their own role in assuring quality in all areas of the firm's activity. TQM and ISO 9000 are concerned with planning to ensure that products and services are effectively carried out by and within the firm. Yet all design firms, and most contractors and sub-contractors, are essentially concerned with, and should be organized on, a project basis. It is thus essential and inevitable that quality assurance is worked and practised on a project by project basis, but save for the terms of the particular contract under which the work is carried out, most, and sometimes all, of the firms have no effective control or ability to create an overall project quality plan. Thus, it is here that the quality management applications in the construction industry change materially from the practices which have developed in manufacturing and other service industries and the concept of total quality and teamwork through partnering must begin as is illustrated diagrammatically in Figs 4(a) and (b).

The project plan must start with the customer/client/building owner/employer and must continue through a unified project management system which embraces in turn the design team, the main contractor and sub-contractors.

The application of second- and third-party auditing also has different implications. An architectural or surveying firm may itself have been the

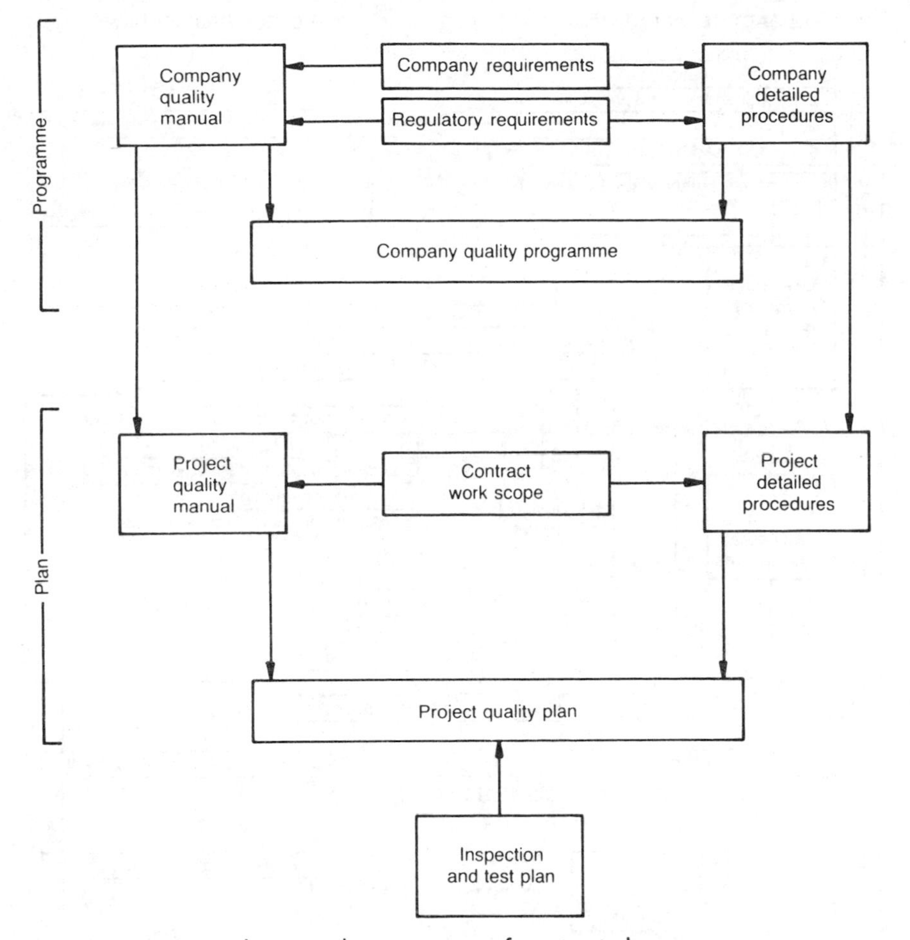

Figure 4(a). Project plan – application to manufacturing industry

subject of second- or third-party auditing and certification and then become second-party auditors in relation to the general contractor (and perhaps also the sub-contractors), whilst the general contractor subjected to auditing by the architect, surveyor or engineer, or a NACCB body, could become the second party auditor or assessor of sub-contractors. But where there are nominated sub-contractors this function presumably has, or should have been, carried out by the employer/building owner as a second party assessor, or by a professional adviser on his or her behalf.

Variations in the type and format of the various contractual relationships which exist and co-exist on a complex building project also have control effects on the various parties. Other controls can also be enforced by public authorities, public utility undertakings and local constraints exercised by adjoining owners through party wall awards, rights of light

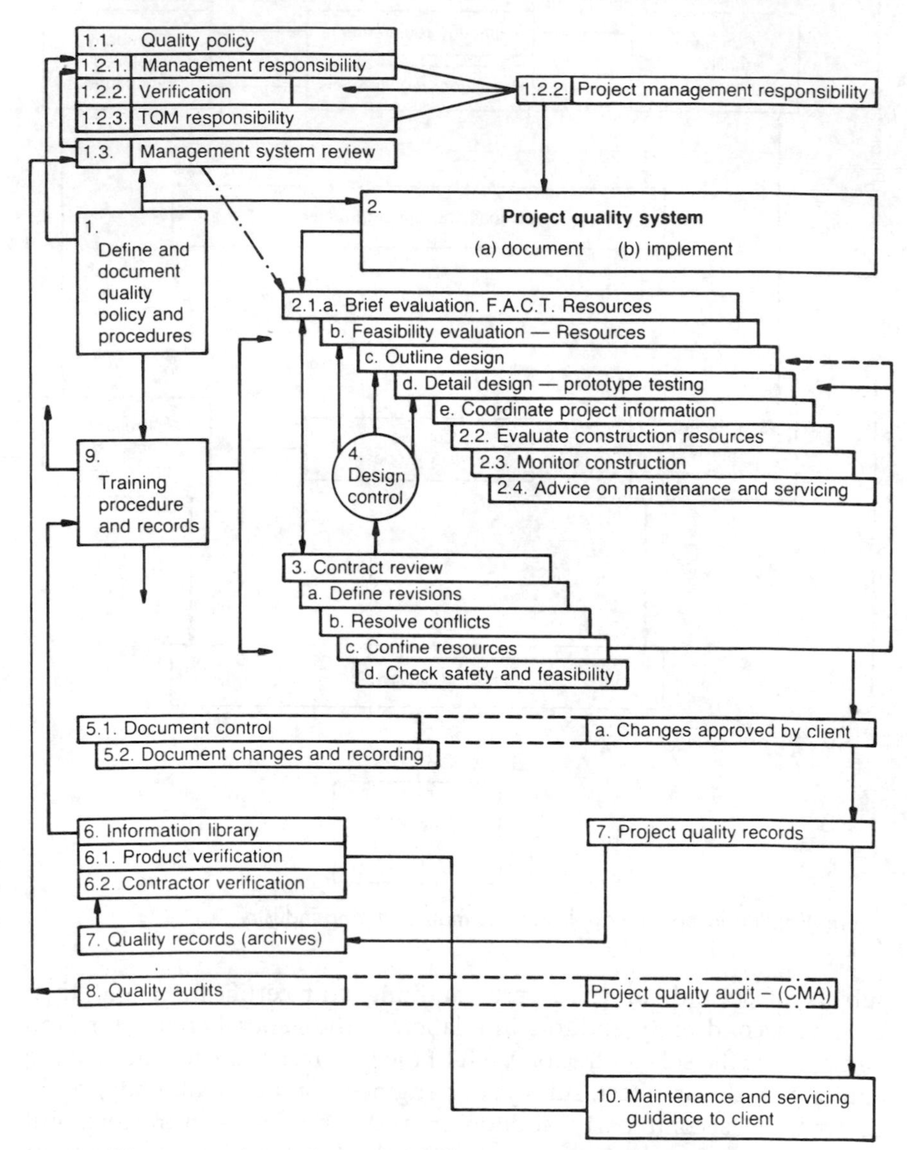

Figure 4(b). Project plan – application to the construction industry

and access, etc. The principles of organization structure and management responsibilities, design and preparation of work procedures which have been developed in individual firms require particular and careful attention on a project basis. This must have regard not only to the communication and control systems and their documentation but also to the nature of and specific requirements for the contractual documents, within which the partnering process functions.

TQM in general contracting firms

The nature of the current construction industry is such that often a general contractor will be carrying out some of the design stages, whether through a design and build contract or some other suitable framework. Therefore many of the implications of quality management for the designers and of the project quality plan will also apply to the general contractor, and much more besides. ISO 9000 deals with many aspects that have comparable applications in the general contractor's organization but others are not so adequately dealt with, as is shown by Table 2.

The UK and EFQM model for self-assessment and the framework for those awards have a much better application in creating TQM within a general contractor's organization. To go further with their consideration in this field would be to get out of proportion the implications on the project and partnering concept. They are dealt with in more depth in Chapters 7 and 8 of *Total Quality in Construction Projects*, including the full manual framework for a typical general contractor's quality system.

The British Quality Foundation Model (UK Quality Award Guide to Self-assessment) provides a depth study of the application of the enablers and results activities as they apply to a general contractor or sub-contractor, no less than to any other organization. The processes of benchmarking for continuous evaluation will be involved but, in particular, contractors will need to examine and create the framework for the following.

Leadership

How well do the executive team and all other managers inspire and drive total quality as the organization's fundamental process for continuous improvement?

A total quality approach in general contractors' organizations should demonstrate visible involvement in leading total quality. Areas to address could include how managers take positive steps to:

- communicate with staff
- act as role models leading by example
- make themselves accessible and listen to staff
- give and receive training
- demonstrate commitment to total quality
- include commitment to and achievement in total quality in appraisal and promotion of staff at all levels

Table 2. Relevance of ISO 9000 series to the contractor's organization

ISO 9000 SERIES / BS 5750 Part 8	Part 1	Part 2	RELEVANCE OF ISO 9000 SERIES TO CONSTRUCTION INDUSTRY FUNCTION/PROCESSES		ARCHITECT	G P SURVEYOR	QUANTITY SURVEYOR	BUILDING SURVEYOR	STRUC/CIVIL ENG	B.SERV/M & E ENG	BUILDING CONTRACTOR	CRAFT SUB-CONTRACTOR	MANUFACURING SUB-CON	
5.2.2	4.1	4.1	Quality Policy & Management Org.		●	●	●	●	●	●	●	●	●	
5.4	4.2	4.2	Quality System		●	●	●	●	●	●	●	●	●	
-	4.3	4.3	Contract Review	☺	●	●	●	●	●	●	●	●	●	Note 1
6.2	4.4	-	Design Control		●	X	X	●	●	●	C		C	
5.4.3.2	4.5	4.4	Document Control	☺	●	X	●	●	●	●	C		C	Nt 1+2
6.2.4.3	4.6	4.5	Purchasing		C	X	C	C	C	C	C	X	X	
-	4.7	4.6	Purchaser Supplied Product		X	X	X	X	X	X	X	X	X	Note 3
-	4.8	4.7	Product Identification	☺	X	X	X	X	X	X	X	X	X	Nt 1+3
6.2	4.9	4.8	Process Control		X	X	X	X	C	C	●	C	●	
-	4.10	4.9	Inspection & Testing	☺	X	X	X	X	X	X	C	C	C	Note 1
6.3.6	4.11	4.10	Inspection/Meas/Test Equipment		X	●	X	X	●	●	●	C	●	
6.3.4	4.12	4.11	Inspection & Test Status		P	P	X	X	P	P	●	C	●	
6.3.5.2	4.13	4.12	Control of Non-conforming Product		P	X	P	P	P	P	●	P	P	
6.3.5.2	4.14	4.13	Corrective Action		P	P	P	P	P	P	●	P	P	
6.2.4.6	4.15	4.14	Handling		X	X	X	X	X	X	●	P	P	Note 3
6.4.1	4.16	4.15	Quality Records		●	●	●	●	●	●	●	●	●	
5.4.4	4.17	4.16	Internal Quality Audits		P	P	P	P	P	P	●		●	
5.3.2	4.18	4.17	Training		●	●	●	●	●	●	●	●	●	
-	4.19	-	Servicing		X	X	X	X	X	X	●	X	●	
6.4.3	4.20	4.18	Statistical Techniques		X	X	X	X	X	X	X	X	X	
					DESIGN						BUILD			

Key
☺ Needs more, or restructuring
● Adequately addressed
X Not so relevant
P Project requirements need expressing differently
C The Contract between customer and suppliers at the different stages should specify requirements, not the Standard.

Notes :

1. The principal context for Quality in Building Design and Construction is on the one-off project, contributed to by many firms, sometimes under separate contracts, or sub-contracts. The product only coming into being at the end of a complex series of processes. With the customer himself much involved but through contractual rather than management control, Project Procedures and a Project Manual, the requirements for which should be specified in the Contract, must be more relevant than the Standard.

2. C - The Contract specifies the requirements. Control is not in the supplier's hands and it is not the "Standard" but the client's specific requirements stated in the Contract which is the relevant criteria.

3. X - In some cases it is not practicable or sensible to deal with the one-off cases in the way described in or envisaged by the Standard.

4. Section 1.1 Scope, states that the requirements (of the Standard) are aimed primarily at detecting non-conformity and preventing its occurrence and it requires the supplier to demonstrate his ability to control the processes that determine the acceptability of the product. In the one-off building project for several reasons, not the least the differences brought about by the contemporaneous forms of contract, the requirements of 9001 and 9002 will NOT provide this assurance.

5. For the design processes a most important requirement is the provision of an adequate Information Database or Library.

6. For both Design and Build a Resource Planning System to assure human resources with appropriate experience will be available at the time required for the project is an essential requirement for project quality.

- recognize and appreciate the efforts and successes of individuals and teams.

How managers are involved in recognition of the efforts and successes of individuals and teams:

- at project, local, section or group level
- at divisional level
- at the level of the organization
- of groups outside the organization, e.g. suppliers or customers.

How managers provide support through:

- helping to define priorities in improvement activities
- funding, learning, facilitation and improvement activities
- actively supporting those taking quality initiatives
- releasing staff to participate in total quality services.

How managers take positive steps to:

- meet, understand and respond to the needs of customers and suppliers
- establish and participate in 'partnership' relationships – with customers and suppliers
- establish and participate in joint improvement activities – with customers and suppliers.

How managers actively promote total quality outside the organization through:

- membership of professional bodies
- publication of booklets, articles
- presentations at conferences and seminars
- support of local community.

Policy and strategy

The organization's mission, values, vision and strategic direction and the manner in which it achieves them.

- How far does the organization's policy and strategy reflect the concept of total quality?
- How have the principles used in the determination, deployment, review and improvement of policy and strategy been successful?

People management

The national standard 'Investors In People' provides a comprehensive framework, as shown on the 'fishbone diagram' (Fig. 5) related to its four principals (Fig. 6).

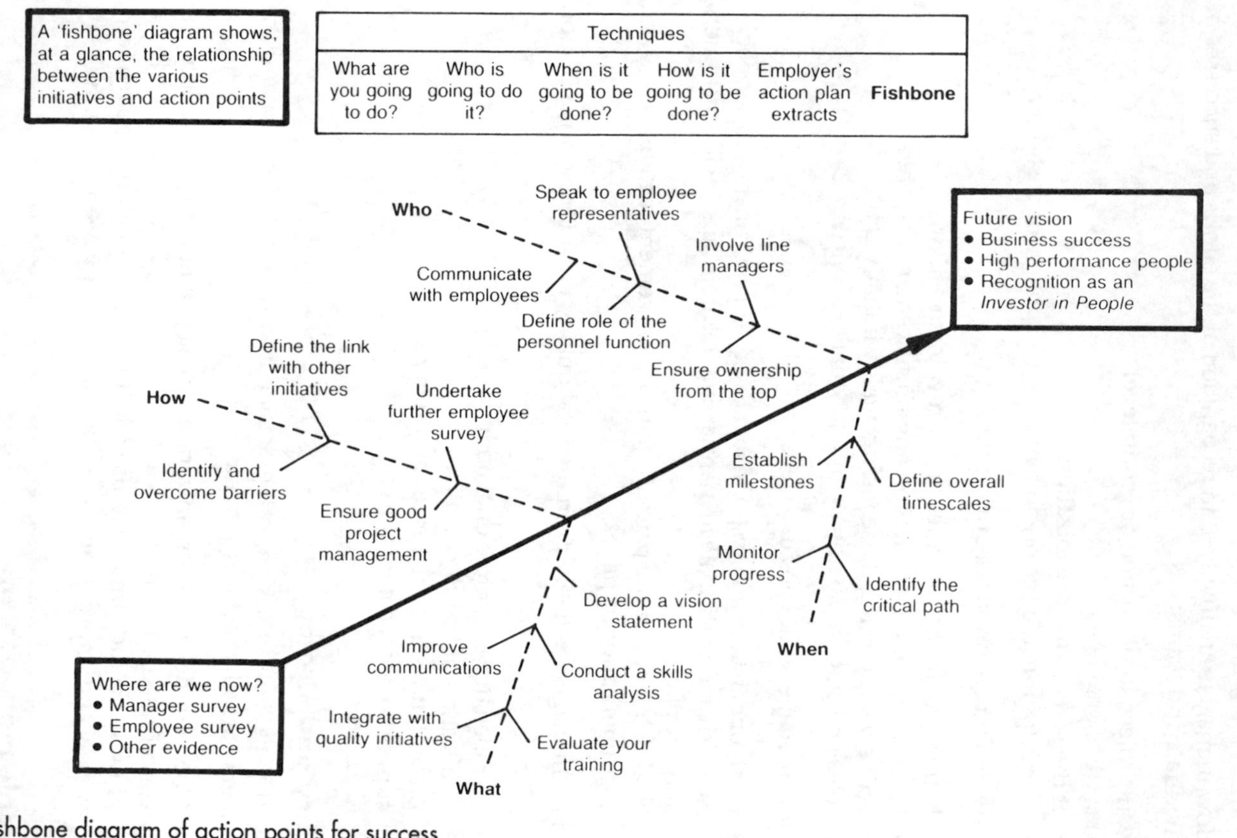

Figure 5. Fishbone diagram of action points for success

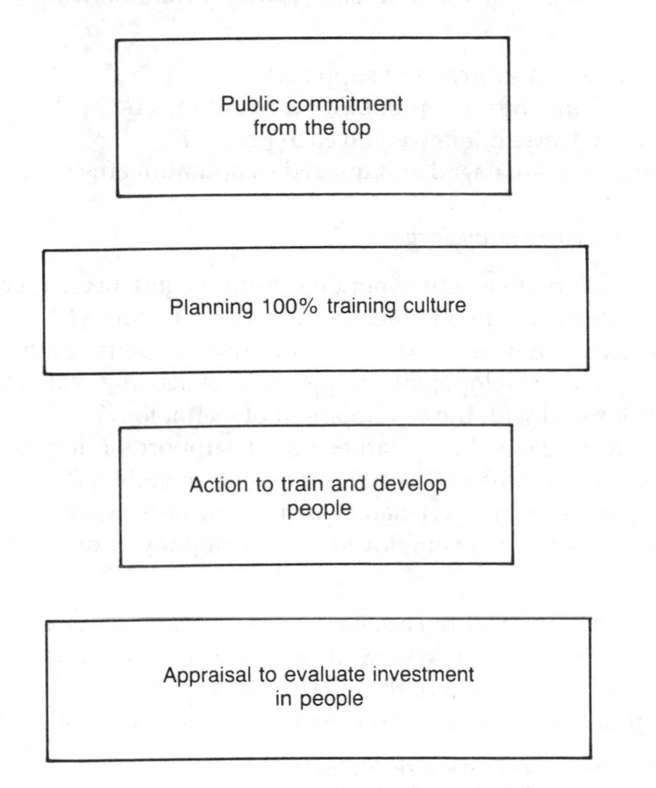

Figure 6. Investors In People national standard – the four principles

Resources

The management, utilization and preservation of resources

Finance. How effectively:

- do financial strategies reflect total quality?
- can they be reviewed and improved?
- can the financial parameters, project budgets, cash flow, balance sheet elements, be managed for improvement?
- can 'quality cost' concepts be used?

Information resources. How can:

- information systems be better managed for improvement?
- information validity, integrity, security and scope be improved?
- information to customers, suppliers and people involved in improvement be made more accessible?
- information strategies support total quality on projects?

Material resources and fixed assets. How can the following be managed for improvement.

- Raw material sources and supplies?
- Material inventories optimized on each project?
- Material waste minimized on each project?
- Fixed assets managed and utilized to optimum effect?

The application of technology.

- Have alternative and emerging technologies been identified and evaluated according to their impact on the business?
- Has technology been exploited to secure competitive advantage?
- Have the development of people's skills and capabilities been harmonized with the development of technology?
- Has technology been harnessed in support of improvements in process and information systems and other systems?
- How does the project nature of the business equate to these aims when many are not employed in the company or division?

Processes

The management of all value-adding activities within the organization. How processes are identified, reviewed and, if necessary, revised to ensure continuous improvement of the organization's business.

Have processes critical to the success of the business been identified?

- How are critical processes defined?
- How is the identification conducted?
- How are interface issues resolved?
- How is the 'impact on the business' evaluated?
- How can these processes be improved on a project basis, when many are not within the division or company's permanent control?

Critical processes include processes associated with the results criteria (BQF Award Criteria 6–9). They include:

- provision of raw materials and supplies
- contract provision
- project management
- pre-tender information and handling
- certification procedures
- detail design
- variation order procedures
- marketing
- estimating
- planning
- management of safety, health, environment

- site supervision
- site security
- sub-contractor investigation
- procurement
- payment of sub-contractors.

Areas to address could include how:

- process ownership and standards of operation are established
- standards are monitored and by whom
- performance measures are used in process management
- quality systems standards, for example ISO 9000, are applied in process management.

How project process performance measures, along with all relevant feedback, are used to review processes and to set targets, including for improvement. Areas to address could include how:

- information is obtained from people, client, architects, engineers, suppliers and competitors
- data developed for benchmarking can be used in setting standards of operation and targets for improvement on a project or divisional basis
- current performance measures and targets for improvement can be related to past achievement
- challenging targets can be identified and agreed upon, either project or region, division or company.

How can we stimulate innovation and creativity in process improvement? For example:

- how new principles of design, new technology and new operating philosophies can be discovered and utilized
- how the creative talents of all employees can be brought to bear
- how organizational structures can be changed to encourage innovation and creativity.

How can the organization implement process changes and evaluate the benefits. For example:

- how new or changed processes can be piloted and implementation controlled
- how process changes can be communicated and to whom
- how staff can be trained prior to implementation.

The score charts, Tables 3 and 4, as used by the assessors for the award illustrate the weightings given to the assessments by the award processes. They are not necessarily the correct weightings for any particular

Table 3. EFQM 'score' charts

CHART 1. THE ENABLERS

The assessor scores each part of the enabler's criteria on the basis of the combination of two factors.

- The degree of excellence of your *approach*
- The degree of *deployment* of your approach

Approach	Score	Deployment
Anecdotal or non-value-adding	0%	Little effective usage
Some evidence of soundly based approaches and prevention based systems Subject to occasional review Some areas of integration into normal operations	25%	Applied to about one-quarter of the potential when considering all relevant areas and activities
Evidence of soundly based systematic approaches and prevention based systems Subject to regular review with respect to business effectiveness. Integration into normal operations and planning well established	50%	Applied to about half the potential when considering all relevant areas and activities
Clear evidence of soundly based systematic approaches and prevention based systems Clear evidence of refinement and improved business effectiveness through review cycles Good integration of approach into normal operations and planning	75%	Applied to about three-quarters of the potential when considering all relevant areas and activities
Clear evidence of soundly based systematic approach and prevention based systems Clear evidence of refinement and improved business effectiveness through review cycles Approach has become totally integrated into normal working patterns Could be used as a role model for other organizations	100%	Applied to full potential in all relevant areas and activities

For both 'approach' and 'deployment', the assessor may choose one of the five levels 0%, 25%, 50%, 75% or 100% as presented in the chart, or interpolate between these values

CHART 2. THE RESULTS

The assessor scores each part of the results criteria, on the basis of the combination of two factors.

- The degree of excellence of your *results*
- The *scope* of your results

Results	Score	Scope
Anecdotal	0%	Results address few relevant areas and activities
Some results show positive trends Some favourable comparisons with own targets in most areas	25%	Results address some relevant areas and activities
Many results show positive trends over at least 3 years Favourable comparisons with own targets in many areas Some comparisons with external organizations Some results are caused by approach	50%	Results address many relevant areas and activities
Most results show strongly positive trends over at least 3 years Favourable comparisons with own targets in most areas Favourable comparisons with external organizations in many areas Many results are caused by approach	75%	Results address most relevant areas and activities
Strongly positive trends in all areas over at least 5 years Excellent comparisons with own targets and external organizations in most areas 'Best in class' in many areas of activity Results are clearly caused by approach Positive indication that leading position will be maintained	100%	Results address all relevant areas and facets of the organization

For both 'results' and 'scope', the assessor may choose one of the five levels 0%, 25%, 50%, 75% or 100% as presented in the chart, or interpolate between these values

Table 4. EFQM 'Score' charts

CHART 3. CRITERIA

Criterion 1 How the executive team and all other managers inspire and drive total quality as the organization's fundamental process for continuous improvement

Part 1a Visible involvement in leading quality management

Areas to address	Strengths
How managers:	Leadership role taken by president
act as role models leading by example	Managers first to attend training then lead training
communicate with staff	Messages reinforced at new product launches
make themselves available and listen to staff	Effectiveness assessed by employee survey
	Areas for improvement
	No integrated process to manage activities
Overall score 60%	Activities not subject to regular review

contracting or sub-contracting firm but do give an indication of the approach that could be used for self-evaluation and thus benchmarking for measuring progress.

It is unlikely that anyone beginning to consider the implications of TQM upon their own organization will be ready to make application for recognition of an award for three years but since the whole approach is aimed at improving the performance the foregoing will be helpful to make the journey at least as interesting and rewarding as the arrival.

The application to a construction project will need first an assessment framework considerably modified (see Chapter 3 and Appendix 3) and then progressive review as the project proceeds.

TQM in contracting and sub-contracting firms

There may be little difference, except in size or organization, between the implications of TQM on a general contractor and on a specialist sub-contractor. This will depend partly on the degree to which the sub-contractor manufacturers standard products which are then installed on-site against specific contractual requirements and partly the degree to which the sub-contractor's role flows from a trade or skill.

TQM is not about paperwork systems, standards and third-party certification but about the creation of an attitude towards improving competitiveness and the performance of the organization as a whole and its ability to provide profitably the service and installed product, or both, that the client requires.

The contractual implications, or perhaps the cultural matters of partnering concerned with the human relations on the project, may have more bearing on and benefit to the sub-contractor, whether of a specialist service, or a specialist manufactured product that must be installed on-site, or a traditional building trade specialist. Partnering will certainly be the key to the holistic approach which must first be brought to the organization and then incorporated into the team performance with other sub-contractors and with the general contractor. That the sub-contractor is indeed regarded by the general contractor as an 'internal customer' will indeed be a major step forward, but nothing short of real demonstration will overcome the current cynicism towards the general contractor and the partnering philosophy. However, the whole concept of partnering must start with the client by demonstration, not invocation, if it is to flow through the project, as it must, from top to bottom.

Without this conviction and its actual demonstration neither the general contractor nor specialist sub-contractor will be able to extend their partnering practice and develop enthusiasm for it amongst their workers, whether on-site or in the office.

3 The partnering concept and construction

I tell this tale, which is strictly true,
Just by way of convincing you
How very little, since things were made,
Thing have altered in the building trade.

A Truthful Song – Rudyard Kipling

The last thing one knows in constructing a work is what to put first.

Blaise Pascal 1623–1662

Quality management techniques and partnering in particular involve philosophies and require changes in culture which come from previous mind-sets. This chapter addresses these as an evolution from TQM into the partnering ambience. It identifies who the project stakeholders are, when and how the partnering process should begin and provides the details of the process at all stages from before the tenders are called for to the project's conclusion. This chapter also looks at the benefits that will result from effective use of the concept and also at the problem areas that could lead to failure. It stresses the importance of top management commitment and good communications and of giving adequate time to the initial partnering workshop. It illustrates examples of successful projects and the charters created for them, and makes reference to Appendix 1 which gives examples of the paperwork created by partnering.

Introduction – team building and partnering in construction

Until total quality of the end project, the building or structure, is delivered to its owner/sponsor the whole quality movement, quality inspection (QI), quality control (QC), quality assurance (QA), TQM, or whatever, means nothing and adds no real value to the construction industry's purpose.

Quality in the building project means providing the client with a building satisfying his Functional and Aesthetic requirements, delivered to the approved contractual Cost and by the agreed Time. Fulfilling the FACTs of the project's brief.

TQM, the philosophies of empowerment, teamwork, Kaizan, benchmarking, or continuous improvement, and the techniques used to develop a repetitive quality product and service to satisfy long-term customers, must be developed in construction by one-off teams and relationships with, often, one-off clients, amid an environment of risk, conflict and potentially adversarial contracts, to produce long-term satisfaction in their use on a single capital project.

As we have seen earlier, the principles and purpose of TQM in other areas and industries undoubtedly will apply and must start, as in the other situations, at the top. But, it will need substantial and careful re-orientation. In the case of building projects the 'top' means with the client, who must at the beginning create the project environment in which the TQM philosophy and techniques can flourish – in a word 'partnering' or, better still in two words 'project partnering'.

General background – conflict and loss

The production process and productivity in the manufacturing process, as well as in many service industries where repetitive activities are involved, recognize that final assembly will use components obtained from other sources.

The development of a good customer/supplier relationship is essential and with it the recognition that both have a vested interest in continuing to trade with each other. Contracts for long-term supply exist and may well have been negotiated at length and in depth about both technical and commercial matters.

Then, if something goes wrong in quality or delivery both parties recognize the benefits of a joint consideration to deal with the complaint, perhaps by a replacement of the faulty goods or correcting the wrong service but without recourse to any penalty clauses that may be written into the contract. Commercial interests and joint financial benefit are readily recognized as more important than the fine points of law enshrined in a simple (or comprehensive) 'buyer' and 'seller' contract.

The construction industry world-wide is frequently confrontational and conflict prone, and has been generally slow to adopt the principles and practices of quality management. Perhaps because the concept originated and was strongly developed within the context of manufacturing industry, and more latterly in repetitive services and processes and the Citizens' Charter for public services. Many of the techniques which have been developed within the context and concepts of total quality management have originated from the original Deming, Juran and Crosby initiatives and so form a strong base of practical statistics and analysis of mass-production.

The people philosophies which have been built into TQM from Deming origins recognize that it is frequently the management system which is at fault and which demotivates people. Improvements have been made through the recognition that people and their motivation are the real keys to the development of successful businesses and service operations.

Delegation or empowerment from top management to the lowest possible level of operation has brought real and very substantial benefits to those managements that have truly adopted total commitment to quality management. Where there is continuity of interest between buyer and seller the success has been even greater. The technique which has evolved is the concept of partnering.

The firm who supplies you and your customer whom you supply are brought into 'partnership' to produce a win/win situation. The philosophy is certainly applicable to construction, for without a truly 'team effort' no construction project will ever be successful. The team in the case of construction must include the client. But it is the client who frequently sets the project off on the wrong foot by seeking to achieve his or her objectives by demanding impossible time schedules, then accepting tenders which are frequently 'loss making' as they stand to contractors, and then by seeking a quality of project far beyond the standards envisaged – and sometimes beyond even those specified.

Not surprisingly, this approach produces an attitude of mistrust, cynicism and a consistent approach to attempt to obtain more than you pay for right down the line. The essentially professional, i.e. disinterested approach, formerly expected from the design team, has been similarly pressured into a catch-as-catch-can commercial merry-go-round extending the jungle of construction into a barrier between, rather than cooperative interface with, the client, and vice versa.

Partnering in the construction industry has, therefore, a much more difficult ambience in which to operate. The first steps must be to overcome this culture of conflict which has evolved ever more strongly over the past 15 years.

Teamwork and profit

Partnering may represent nothing more than a return to good relations, honesty, integrity and cooperation which has been the hallmark of good building in the past and, indeed, even in the present century, but it works.

Cultural changes are clearly needed and a clear direction to this end runs through the reports of Sir Michael Latham, 'Trust and Money' and 'Constructing the Team' (HMSO, July 1994).

Partnering has been shown to work – is working – in the United States and in Australia, and on projects where cooperation and collaboration have been instituted from the commencement. In some cases this has been consciously with an awareness of the philosophical and practical benefits that came from quality management, in others from organizations which almost intuitively just set about developing a climate of consent and agreement.

Partnering is choosing to live by the spirit rather than the letter of the law, values treasured in an ethical democratic society. Litigation is not counter-productive when it defines both legal and factual issues, but that is often not the case. Lawsuits are inimical to the basic nature and goal of the industry. Construction is not an individual endeavour like long-distance running but rather a business of team-building. The fabric of the industry depends on strong weaving of owner, architect, engineer and contractor into a team. Successful teams are built on the strengths of each member, while successful lawsuits are founded on capitalizing on the weaknesses of team members.

It is parties to a contract, not their lawyers, who must decide whether they want 'tough' contracts harshly applied or management that envisages good relationships and teamwork where everyone wins. This is what it is at stake.

Teamwork should involve all the stakeholders

The stakeholders in the construction project partnering activity are therefore:

- building owner (client) – and his/her financiers
- design team
- main contractors
- specialist contractors
- sub (sub)-contractors
- major suppliers

all of whom stand to benefit from partnering.

Clearly the technique can more easily be developed on term contracts between organizations with a continuing need for a construction contractor's services where the client's team and contractor's team can develop the culture and time and cost-saving procedures to produce these benefits against operating benchmarks from task to task and time to time. But, partnering does not eliminate competition either locally on one-off projects or on a world stage, for as Lester C. Thurow* in his forecast of the coming economic battle between Japan and the Pacific Rim, America and Europe, says:

*Professor of Economics and Dean of the Sloan School of Management in M.I.T. in *Head to Head*, published by William Morrow & Co., Inc., 1992.

'In the intense competitive atmosphere that will exist in the 21st century all the participants should remind themselves daily that they play in a competitive – cooperative game, not just a competitive game. Everybody wants to win but co-operation is also necessary if the game is to be played at all'.

Thurow reviews some strategic factors which characterized the present century, taking into account the collapse of the Soviet Union and Eastern Europe and the reality of the European Community and believes that in a sense the 21st century has begun already – and is already different from the last. Thurow thinks historians will label the 19th as the century of Great Britain and the 20th as that of the United States; then asks to whom will belong the 21st?

Thurow believes: '. . . without pause the contest has shifted from being a military contest to being an economic contest – in which – the world is not divided into friend and foe. The game is simultaneously competitive and cooperative. One can remain friends and allies, yet still want to win'. On the macro scale he identifies three protagonists: Europe, Japan (and the Pacific rim of south-east Asia), and the United States, and makes a closely argued case for each as world leaders. The combination of the thinking of American consultants with the Japanese culture has certainly blazed the trail for TQM and the cooperative approach of partnering, albeit in a mass productive environment.

The process

But partnering can, and has, worked successfully on one-off projects in Japan, the United States and Australia, maybe for different background reasons, but giving the will and the understanding of the principles and key elements to:

- create partnership charter
- start at the top
- create mutual objectives, goals and win/win thinking – equity involvement
- develop trust and teamwork (eliminate cynicism)
- implement joint strategies and problem solving mechanism
- empower personnel
- joint review and evaluation
- create rapid response and problem resolution.

It may be, too, that the fundamental reasons for the cultural background of Japan which have led to the successful exploitation of those management concepts are closer to the fundamental reasons which have developed the UK's construction culture and so will help its acceptance here once the benefits can be demonstrated.

Successful introduction will produce the benefits to all – even in the UK's hostile confrontational pre-Latham environment!

And for the design consultants there will also be reduced exposure to litigation and to document deficiencies. It will give them an enhanced role in problem solution and decision-making, lower administrative costs and greater opportunity for profit if they have the equity involvement through the partnering mechanism.

So, what is the process?

The partnering concept is not a new way of doing business – some always conduct themselves in this manner. It is going back to the way people used to do business when a person's word was their bond and people accepted responsibility. Partnering is not a contract, but a recognition that every contract includes an implied covenant of good faith.

The contract establishes the legal relationships, the partnering process attempts to establish working relationships among the stakeholders through a mutually-developed, formal strategy of commitment and communication. It attempts to create an environment where trust and teamwork prevent disputes, foster a cooperative bond to everyone's benefit, and facilitate the completion of a successful project. It does, of course, require a process of change in attitude, a subject we will explore in depth in Chapter 5. Change of any kind requires both education for and training in the 'why' and the 'how' and 'who'.

The 'who' in this case is a top-to-bottom and side-to-side involvement of the on-site project team, and many of their off-site colleagues and dependencies. All will need to be put through the four phases of change illustrated in Fig. 7.

For the most effective results, stakeholders should conduct a partnering workshop at the early stages of the contract. The sole agenda of the workshop is to establish and begin implementing the partnering process. This forum produces the opportunity to initiate the key elements of partnering.

They are as follows.

Commitment

Commitment to partnering must come from top management. The jointly developed partnership charter is not a contract but a symbol of commitment within it. When the commitment is made it should be strongly and widely communicated to all the stakeholders and their employees and the whole project community (see Figs 8 and 9 for US and Australian examples).

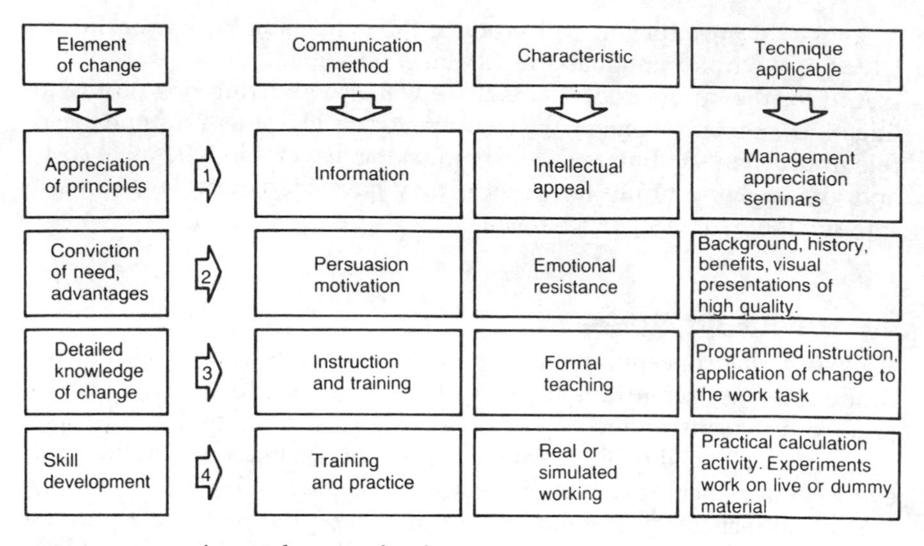

Figure 7. Four phases of training for change

Equity

All stakeholders' interests are considered in creating mutual goals and there is commitment to satisfying each stakeholder's requirements for a successful project by utilizing win/win thinking. Where many sub-contractors are involved, some at later stages in the project, they should at least be made aware of the initial partnering workshop and then attend another specially convened at the later stage to ensure they are fully involved in the new project culture.

Trust

Teamwork is not possible where there is cynicism about others' motives. Throughout the development of personal relationships and communication about each stakeholder's risks and goals, there is better understanding. With understanding comes trust and with trust comes the possibility for synergy.

Development of mutual goals/objectives

At a partnering workshop the stakeholders identify all respective goals for the project in which their interests overlap. These jointly-developed and mutually-agreed-to goals may include achieving value engineering savings, meeting the financial goals of each party, limiting cost growth, limiting review periods for contract submittals, early completion, no lost time because of injuries, minimizing paperwork generated for the purpose of case building or posturing, no litigation, or other goals specific to the nature of the project and the parties' aims.

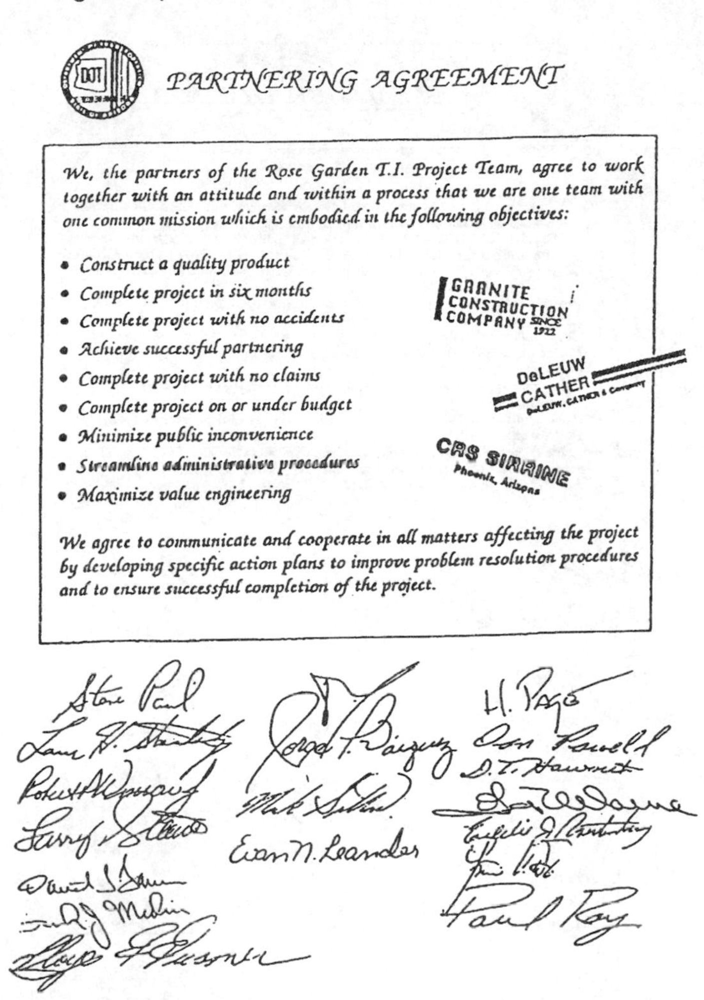
PARTNERING AGREEMENT

We, the partners of the Rose Garden T.I. Project Team, agree to work together with an attitude and within a process that we are one team with one common mission which is embodied in the following objectives:

- Construct a quality product
- Complete project in six months
- Complete project with no accidents
- Achieve successful partnering
- Complete project with no claims
- Complete project on or under budget
- Minimize public inconvenience
- Streamline administrative procedures
- Maximize value engineering

We agree to communicate and cooperate in all matters affecting the project by developing specific action plans to improve problem resolution procedures and to ensure successful completion of the project.

Figure 8. The partnering charter for the Rose Garden TI project, Arizona, USA

Implementation

At the workshop stakeholders together develop strategies for implementing their mutual goals and the mechanisms for solving problems.

Continuous evaluation

In order to ensure implementation, the stakeholders should also agree to a plan for periodic joint evaluation based on the mutually agreed goals to ensure the plan is proceeding as intended and that all stakeholders are carrying their share of the load.

DAIKYO

GREEN ISLAND RESORT PROJECT
CAIRNS • AUSTRALIA

We, the partners in the Green Island Resort Project, will complete this unique eco tourism development on an environmentally sensitive coral cay within the Great Barrier Reef Marine Park, which will achieve acclaim from our client, the public and within our industries. We are committed to implementing the Partnering Process throughout this project, with the following objectives:

- **Provide interesting, enjoyable and satisfying work for all.**
- **Achieve or better program targets.**
- **Carry out a safe job.**
- **Maintain an effective ongoing team relationship.**
- **Generate future work.**
- **Carry out appropriate, timely and effective verbal and written communication.**
- **Resolve all issues at the lowest level in a timely and progressive way.**
- **Build a quality job.**
- **Achieve community and environmental expectations.**
- **Minimise the impact on construction and tourist operations.**
- **Achieve all parties' financial goals.**

Figure 9. The partnering charter for the Green Island Resort project, Queensland, Australia

Timely responsiveness

Timely communication and decision making not only saves money, but also can keep a problem from growing into a dispute. In the partnering workshop the stakeholders develop mechanisms for encouraging rapid issue resolution, including the escalation of unresolved issues to the next level of management or contract management adjudication.

Benefits of partnering

For all the stakeholders of a project, partnering is a high leveraged effort. It may increase staff and management time up front, but the benefits accrue in a more harmonious, less confrontational process, and at completion in a successful project without litigation and claims.

The partnering process empowers the project personnel of all stakeholders with the freedom and authority to accept responsibility to do their jobs by encouraging decision making and problem solving at the lowest possible level of authority. It encourages everyone to take pride in their efforts to dispense with cynicism and so get along with each other better. The partnering workshop is a particular opportunity for public sector contracting, where the open competitive-bid process has kept the parties at arm's length prior to award, to achieve some of the benefits of closer personal contact which are possible in other situations.

It is interesting to note that the list of benefits to the various stakeholders confirm the mutuality of their individual interests (Table 5).

Potential problems

Partnering requires that all stakeholders 'buy into' the concept. The concept is endangered if there is not true commitment.

- Those conditioned to the adversarial environment may be uncomfortable with the perceived risk in trusting.
- Top management giving just lip-service to the term; treating the concept as a fad is not true commitment at any level.
- For some, changing the myopic thinking that it is necessary to win every battle, every day, at the other stakeholders' expense will be very difficult. Win/win thinking is an essential element for success in this process.
- Not bringing in all the key players, e.g. sub-contractors, at an early stage.
- Skimping the early activities, or the workshops. Changing cultures is not easy.
- Using a standard approach and not tailor-making to the project – and the people on the project.

The partnering process

The following is only a model as illustrated in Fig. 10. Since every project is unique and the particular stakeholders for each project will vary, the process should be tailored by and for these stakeholders for the project. A partnering process can be developed for any type of project and any size project. For the small project the differences will be in intensity.

Developing the process

Step 1. Educate your organization. As with many other benefits they will

Table 5. Table of benefits to stakeholders

Conception and design phase	
Benefits to project owner	**Benefits to design consultants, project architect/engineer and consultants**
Reduced exposure to litigation	Reduced exposure to litigation
Lower risk of overrun – time – cost Better quality product	Reduced exposure to document deficiencies by early identification of problems and their resolution.
Potential for time and cost reduction Efficient resolution of problems	Enhanced role in decision-making and problem solutions.
Lower administrative costs	Reduced administrative costs eliminating defensive case building
Better opportunity for innovation (value engineering) Greater opportunity for financial success	Greater opportunity for profit through equity involvement as stakeholder
Implementation phase	
Benefits to project contractor	**Benefits to sub-contractors**
Reduced exposure to litigation	Reduced exposure to litigation
Better productivity	Improved cash flow under contract
Expedited client/design decisions	Equity involvement improves
Better overall time and cost control	opportunity for innovation and value engineering
Lower risk of over-run – time – cost	Involvement in decision making avoids costly claims
Lower overhead costs	Reduced overhead costs
Greater profit potential	Better and more reliable programming Greater profit potential
Enhanced repeat business	Enhanced opportunity for repeat business

only be achieved if the right involvement is made. Whether you are an owner or a contractor you must educate your own organization about partnering before attempting a project using the concept. Understanding and commitment are essential.

Step 2. Make partnering intentions clear. The owner's intention to

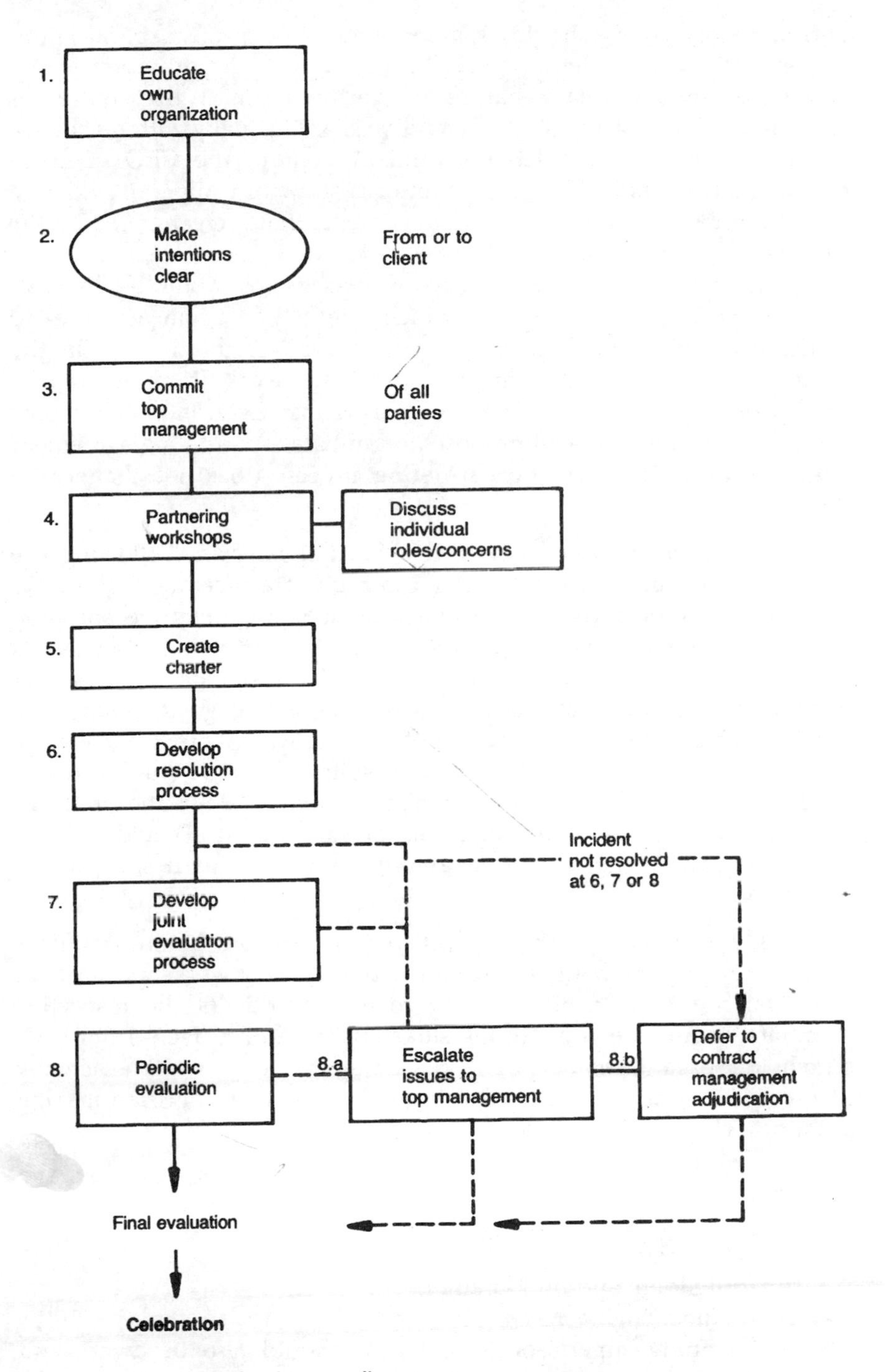

Figure 10. Flow chart for the overall partnering process

encourage partnering should be mentioned in the project specifications and invitations to tender. The specification and tender provision would emphasize the voluntary nature of partnering and that the costs associated with implementing it would be shared equally with no change in the contract price. A letter should be sent to the CEO of every company on the bid list. A sample letter is shown in Appendix 1, along with a number of other forms and documents that are generated by the process.

Second party auditing as a process of pre-selection would be the ideal time to open up the dialogue – which might then begin at a pre-bid conference which could include a presentation on partnering. In the context of a negotiated contract for private work it might be the contractor who proposes the use of partnering. Even in public works contracts the contractor can propose and initiate a partnering agreement after the award because the partnering process does not change the contract.

Step 3. Top management must be committed at the start. Following the award, the owner or the contractor can request a meeting at the CEO level to discuss the partnering approach to managing the project and the CEO role. Commitment at this level is essential for partnering to achieve its potential. Upon agreement, each entity will designate a partnering leader. These leaders will meet at a neutral site to get to know one another as individuals and to plan a partnering workshop. If not already involved, that is the point at which a consultant could be employed as a facilitator – and if the consultant is also to have a role in the problem-resolution process as adjudicator, his or her presence could help in overcoming any vestigial cynicism amongst any of the parties, e.g. a sub-contractor.

Step 4. The partnering workshop. As soon as possible, before problems arise, all key players should participate in a partnering workshop, again at a neutral site away from the job site and outside of the respective corporate cultures of the various stakeholders. Key players from each stakeholder organization at the workshop are those who will be actually involved in contract performance and those with decision-making authority. They might include:

- the owner's manager or representative
- the contractor's area manager
- project manager
- superintendent and project engineer
- the architect/engineer's chief designer
- the planning supervisor (CONDAM) should also be aware and involved

- construction administrator and consultants
- sub-contractors' project managers or superintendents

and, depending on the project, special participants such as a representative from a testing laboratory or a key public official.

A facilitator is not essential to the concept. The designated partnering leaders should have initiated this workshop and they are the ones in charge of it. They should introduce the partnering concept and outline the objectives to be developed in the workshop. The facilitator could be used to develop the background knowledge to ensure all parties are on an equal footing of knowledge.

While partnering workshops are most effective at the beginning of a new project, project relations and problem-solving can be improved even midway on a project, or at key stages of progress.

Creation of the partnering charter. Through the identification of stakeholders' respective goals for the project, mutual objectives can emerge. These mutually developed objectives form the partnering charter. In this process the players get to know one another and develop team attitudes rather than we/they attitudes. The facilitator can play a very useful role in helping to establish these goals and getting real commitment to them earlier – rather than later. Partnering goals might include:

- achieving specific value engineering savings
- limiting cost growth
- limiting review periods for contract submittals
- early completion
- no lost time because of injuries
- minimizing paperwork generated for the purpose of case building or posturing
- no litigation
- other goals specific to the nature of the project.

Making the construction project an enjoyable process may be part of the charter. The charter not only is a symbol of the stakeholders' commitment to partnering, but also can be used as the scale against which the stakeholders' implementation of the process can be evaluated. The ceremonial signing of the charter after the personal interaction necessary for the development of mutual goals is an important formalization of the bonds among all stakeholders. Sample charters from the United States and Australia were shown earlier in Figs 8 and 9.

Development of an issue resolution process. In the workshop the key players should design their own systems for resolving issues on the project. Specific teams composed of personnel from the various stakeholders, who are knowledgeable about their particular technical portions of the contract, discuss potential problems and the way they would like to see them handled. They decide how issues that are not resolved at their level will be escalated to the next level in a timely fashion so that the decision-making process becomes more efficient and costly delays are avoided. If the contract (post-Latham) contains no provision for adjudication this should be incorporated. This is dealt with in depth in Chapter 7.

Development of a joint evaluation process. In partnering the effectiveness of the process participation is reviewed and evaluated periodically by all participants – not just evaluation by the contractor or the owner. Evaluation can be in periodic written form, through periodic meetings of the key players, and periodic executive meetings. Evaluation, of course, includes recognition of positive behaviour and not just deficiencies.

Discussion of individual roles/concerns. Workshop discussion should include definitions of each key player's unique role and what makes the job successful for that role – what that individual needs and how it is needed.

A workshop goal should be a high-trust culture in which everyone feels they can express their ideas and contribute to the solution.

Risks and potentially difficult areas of the contract should be discussed openly. Players should be made aware of the potential for value engineering. Understanding other stakeholders' risks and concerns and seeing where one's portion of the contract fits in relation to others' helps to build the essential team attitude. In the process, individuals grow to know and understand the personalities with which they will be working before problems have arisen. This investment in the human dimension at this point can reap very significant benefits for the life of the project.

The facilitated workshop. The neutral facilitator can be very helpful in organizing the workshop agenda and providing training in conflict management, listening and communication skills, as well as insights into individual problem-solving styles. The role of the facilitator, however, is not to lead, but to keep the focus on, and improve, the process, to elicit from the stakeholders what they want out of the workshop, as well as their particular goals and objectives for the project. All stakeholders must be comfortable with the facilitator.

If the facilitator is the adjudicator under the contract this is even more important, it will also give him or her an insight into the project and its potential problem areas, and this will have included the personality and perhaps management styles of the characters involved. If and when matters are later referred to the facilitiator while in the role of adjudicator then he or she will be able to 'blow the whistle' more quickly, make an award, and get the game going again quickly.

The facilitator can help produce the products: the partnering charter, the issue resolution system and the joint evaluation system. A good facilitator would be neutral but with some understanding of construction. The facilitator should also obviously have organizational skills.

Figure 11. Preserving the relationship?

A qualified facilitator is particularly valuable in initial partnering experiences to help develop comfort and confidence with regard to the effective implementation of the partnering process. The benefits of using a facilitator should be balanced in the light of the facilitator's fee in relation to the total cost of the project. The long-term advantage of the training for each organization's personnel should not be overlooked.

A typical workshop agenda and exercises that could be tackled during the workshops is given in Appendix 1, pages 169 and 170.

Periodic evaluation

Formal, periodic, evaluation helps ensure that the relationships and attitudes created in the workshop are not lost. It also helps to keep the project implementation on target by looking back at goals and assessing progress in relation to those goals. Sample written evaluation forms are given in Appendix 1.

Occasional escalation of an issue

Conflicts are inevitable in any human endeavour. Key players should be encouraged to escalate to the next level of management the issues they are unable to resolve themselves. Escalation saves time and money. It may prevent the stakeholders from taking a rigid position and thus keep a relatively minor issue from becoming a claim. Most importantly, it may preserve the working relationship of the key players. When otherwise the conventional tendency for disputes to arise and the self-destruct button become dominant. Timely, effective outside help will enable the parties to live to fight another day – but together towards common objectives.

Table 6 is an agenda for a simple one-day partnering workshop. For larger projects the parties may wish to expand the time and scope of the workshop by including discussions of problem solving styles, prior experiences, risk management philosophies, anticipated difficulties, and/

Table 6. Typical partnering workshop agenda

9.00–9.15 a.m.	Opening remarks of senior executives – why we are here. Client – contractor – others
9.15–9.30 a.m.	Introductions
9.30–10.30 a.m.	Partnering overview (by facilitator/project manager/ employer's representative?)
10.30–10.45 a.m.	Break
10.45–11.15 a.m.	Exercise No. 1 Barriers, problems, opportunities
11.15–11.45 a.m.	Report and discussion in entire group
11.45–12.00 noon	Develop project mission statement
12.00–1.00 p.m.	Lunch
1.00–1.15 p.m.	Exercise No. 2 Interests, goals, objectives
1.45–2.15 p.m.	Report, discussion, identification of common goals and objectives
2.15–2.30 p.m.	Break
2.30–3.15 p.m.	Exercise No. 3 Issue resolution and team evaluation
3.15–4.00 p.m.	Report discussion, agree on process and format
4.00 p.m.	Sign charter

or simply give more time for the parties to become better acquainted – in small or larger groups.

Final evaluations and celebration

Final evaluations are a way of learning from the experiences of the project. Closure and celebration are important human considerations. The partnering process could have changed mind-sets. It should have, at worst, allowed energies to be focused on de-confliction rather than confliction.* Success should be recorded and made known to all involved. Partial failures should also be recorded and, if possible, quantified as a basis for targets for improvement on future work on future projects. For some time these benchmarks will be valuable motivation for future projects – your own and others!

Partnering has the potential to change our industry, one project at a time.

***Confliction* is word created and defined by Edward de Bono as the process of setting up, promoting or encouraging conflict. It is the deliberate effort put into creating a conflict. *De-confliction* is the opposite of confliction. Designing away or dissipating the basis of conflict. The effort required to evaporate a conflict.

4 Pre-partnering qualification and selection

He who has begun has half done.

Have the courage to be wise.

Horace 65BC–8AD

Before partnering is presented to the various parties involved some attempt at pre-selection should be carried out of only those likely to be sensitive to the specific project's need for their services and the culture that will be necessary for fully integrated teamwork.

This second-party auditing is described and the purpose of the questionnaires needed and their role in preparing the would-be partner for the project are discussed with the specific questionnaire forms included in Appendix 1.

For some years to come the concept of partnering will be new to many undertaking construction projects. This will certainly apply to clients and to their design teams and to the smaller general contractors and sub-contractors.

The selection of these firms as a pre-qualification for tendering will therefore take on a new and added importance. How this will be approached will depend upon the context of the project and, more particularly, where the initiative for the partnering concept originates.

For some time it has been the practice of the better contractors to investigate sub-contractors for sub-contract packages and many will have preferred sub-contractors and information on others which they use for both formal or informal enquiries prior to tendering. This, in the language of ISO 9000 is closely akin to second-party auditing but with the audit prescription that it must be project-specific.

Included in Appendix 1 is questionnaire B for second-party project auditing and project management as a pre-partnering qualification and aid to selection. The three questionnaires:

A. for architectural/engineering/surveying organizations
B. for the general contractors
C. for specialist sub-contractors

are designed to be used in the three main situations for which second-party audits will be undertaken related to a specific project's resourcing needs. They illustrate the essential maxim for quality management – that the philosophy, commitment, attitudes, practices and procedures must start at the top, and only then will they permeate down through all levels and areas of the project organization.

As we have seen, for the construction project the top is, and must be, the client for the project and it is the client who will become the second party to all contracts entered into. First with the design team, next with a main contractor or contractors and sometimes with sub-contractors for specialist work.

The client may appoint a project manager either from within his or her own organization – or as a further contractual relationship for which a second-party audit is desirable, if not even more essential. But, it is the client who must determine:

- the key factors for the project
- what the project is
- where and when it is to be built
- to what standard and order of cost it should be built
- who will be selected to carry out the work.

The client may delegate much, or even all, of the work to second-party auditors but the decisions remain his or hers (see Fig. 12 for timing).

The prime purpose of these pre-qualification auditing procedures is to see that the organization selected to carry out the project, or their part of it, is itself equipped to and can provide a reasonable assurance that their resources and management will meet the project's requirements and client-defined standards in every way. This is not just a matter of documentary procedures but also of people systems and perhaps their motivation, and sensitivity to a partnering project. The questionnaires, included complete as models in Appendix 1, seek to establish that both exist. They should be introduced in a manner suited to the contract in which each will be used.* For example for the following situation.

Client/design team professionals

'The purpose of this project questionnaire is to establish the comprehensive effectiveness of your management methods, systems and people to

*Taken from *2nd Party Project Auditing*, Polycon, 1993.

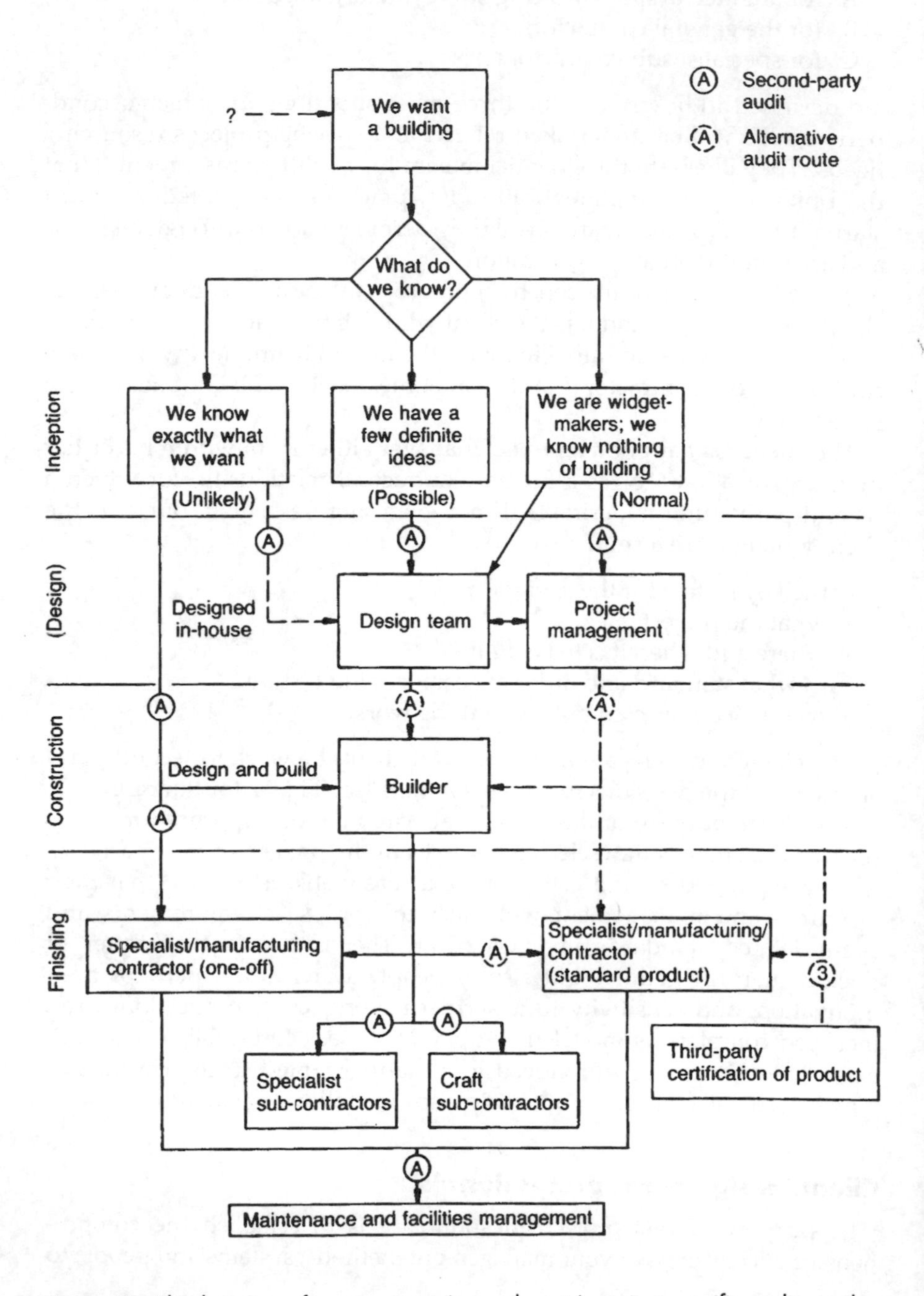

Figure 12. The location of pre-contract (second-party) project-specific audits in the client alternative route through the building process

'get it right first time' economically and effectively. It will also have regard to your company's contracting and legal procedures with particular emphasis on measures to prevent contractual disputes arising and resolving them quickly when they do.

This audit is concerned with the totality of the factors that affect your company's quality management and could therefore be said to be of its 'quality management system', i.e. *the organization structure, responsibilities, activities and resources and events that, together, provide organized procedures and methods of implementation to ensure the capability of your firm to meet our specific project quality requirements.*

Our aim is to develop a 'project quality plan' which is a documented set of activities, responsibilities and events serving to implement the project quality system. The quality plan must of necessity embrace all those involved in the project from the time of commencement. It will therefore include the work of the design team and their interface with contractor and sub-contractor. It must also embrace the activities, operation and decisions made by or on behalf of 'the client'.

This second-party audit, whether carried out by a consultant or by a project manager, is aimed at:

(*a*) identifying the objectives and scope of the systems to be established
(*b*) determining the present stage of development of your quality procedures and their documentation
(*c*) providing a report for discussion by the quality policy or steering group, which will also provide the basis for the project quality programme leading up to final assessment / selection
(*d*) introducing the concept and philosophy of partnering which is to be used on the project.

This questionnaire is intended to help in the selection of either professional organizations, or a contractor/sub-contractor, through whom the client wishes to carry out development or building works. This audit will always be limited to those aspects of the audited company relevant to the client's specific project needs.

Many matters involved in the project quality system will affect more than one department or area of management responsibility and the questionnaire is intended to provide information that will aid selection. It is likely that the person receiving this may not be familiar with all aspects of his or her company, and so the questionnaire will form a good checklist for him or her also.

The questionnaire is therefore divided into sections which may have some elements of overlap where activities are not primarily the responsibility of only one function and of the management related to this function. Each questionnaire has a section 'general organization' for completion by the managing director, senior partner, etc.; also sections dealing with marketing, personnel, industrial relations and safety, resource planning and finance, defects and general claims'.

Sub-contractor

At the time of approaching a sub-contractor it is likely that the project has been progressed to the stage when main decisions have been made, detailed design has progressed, and the early stakeholders will have been established. Therefore, whether the approach to the sub-contractor is by the client or the client's representative, a member of the design team, or the general contractor, rather more information can be sent to the company prior to his second-party audit than has been given in similar form to the design team and then the main contractor. This is included in Appendix 1 in front of the general questionnaire for sub-contractors.

The sub-contractor will, however, also need to be given something by way of introduction, for example:

'The purpose of the project questionnaire is to establish the comprehensive effectiveness of your management methods, systems and people to 'get it right first time' economically and effectively. It will also have regard to your company's contracting and legal procedures with particular emphasis on measures to prevent contractual disputes arising and to resolve them quickly, economically and within a win/win framework of partnering.

'This audit is concerned with the totality of the factors that affect your company's quality management and could therefore be said to be of its 'quality management system', i.e. *the organization structure, responsibilities, activities and resources and events that, together, provide organized procedures and methods of implementation to ensure the capability of your firm to meet our specific project quality requirements.*

'Our aim is to develop a 'project quality plan' which is a documented set of activities, responsibilities and events serving to implement its quality system. The quality plan must of necessity embrace all those involved in the project from the time of commencement of the development. It will therefore include the work of the design team and the interface with contractor and sub-contractor. It must also embrace the activities, operation and decisions made by or on behalf of 'the client'.

Audits similar to this and with the same purpose of effective team building have been carried out on the design team and all contractors and principal sub-contractors. This second-party audit, whether carried out by a Polycon consultant or by a project manager, is aimed at:

(*a*) identifying the objectives and scope of the systems to be established
(*b*) determining the present stage of development of your quality procedures and their documentation

(*c*) providing a report for discussion by the quality policy or steering group, which will also provide the basis for the project quality programme leading up to final assessment/ selection
(*d*) including the concept, philosophy and processes of partnering being used on that project. An initial partnering workshop will be held to further develop this concept.

'This questionnaire is intended to help in the selection of either professional organizations, or a contractor/sub-contractor, through whom the client wishes to carry out development or building works. This audit will always be limited to those aspects of the audited company relevant to the client's specific project needs.

Many matters involved in the project quality system will affect more than one department or area of management responsibility and the questionnaire is intended to provide information that will aid selection. It is possible that the person receiving this may not be familiar with all aspects of the company, and so the questionnaire will form a good checklist for him or her also.

This questionnaire is therefore divided into sections which may have some elements of overlap where activities are not primarily the responsibility of one function and of the management related to this function.'.

The questionnaire will have additional sections over those used for the design team covering contract management, cost control, buying, dispute resolution, whether there are subsidiary companies, whether they manufacture in whole or in part, or whether there is any off-site assembly'.

The completion of these second-party auditing questionnaires and their return to the initial stakeholder can allow a first selection to be made and therefore follow-up auditing visits would be limited to those organizations which appear to have the potential for partnering on the project. At the audit visit these 'human systems' and cultural questions should receive most of the attention.

The sub-contractor, particularly if providing one of the traditional construction trades, will probably need more convincing than others that the client, and general contractor, are really committed to partnering, and are not just using it as one more ploy to beat the sub-contractor down on price or time.

The project specific audit visit in a partnering project

The particular purpose of the visit following completion of a questionnaire is to examine the systems of the company with particular regard to the project objectives laid down in the brief, which will include the partnering intention and to good managerial practice. The particular location to be visited will have been established as that likely to be concerned with the specific project. Before the auditor visits the company he or she will have reviewed the replies to the questionnaire. This analysis will be in relation to the overall project brief and the client's strategy for the specific project under review. Concern for quality will have begun from inception of the project and continue to its final conclusion, that is commissioning and occupation. The client and the design team and general contractor will be fully aware of the extent of their own likely involvement and of the benefits which they hope to obtain in the quality and lower cost of the project construction.

Procedure

The visit of the auditor should start with discussion with a director, principal, or senior partner along with anyone else who filled in the questionnaire. The reasons for the audit will be clearly established with the senior members of the audited company. The sub-contractor should be informed on the first occasion that it is part of a selection process and that the auditor's report will make clear recommendations to the commissioning client. If more than one company is audited against a single requirement the relative advantages and disadvantages, strengths and weaknesses, of the audited companies will be tabulated in relation to the project so that the auditor's client can make an informed decision.

Ultimately this decision may be made on a competitive quotation for price or time. The audit should, however, have established the reliability and quality factors. The format of the assessor's report will clearly vary depending upon whether the audit is carried out before, or after a comprehensive bid has been obtained. Unless specifically so instructed the assessor's job is *not* to make the selection but to assess the criteria to enable the client to do so, and decide on the sensitivity of the sub-contractor to the partnering culture.

It is necessary to confirm any impressions given in replies to the questionnaire regarding work currently being undertaken and the resources committed to such work. Part of the auditor's task will be to examine carefully examples of previous or current work undertaken by the company to assess quality and other aspects relative to the project which is the subject of the audit. Throughout the audit the distinction between how the company as a whole is run and what procedures come into play for individual projects will be made. The likely management structure for the project under review together with the c.v.s (or pen pictures) of the senior personnel should be clearly established, and any additional current workloads for those individuals noted, even though in a small sub-contracting firm there may be little difference.

In companies where a method statement and a programme have been submitted as part of the questionnaire response, the auditor must ensure through discussion, that he or she is satisfied with such statements and that they broadly concur with the client's brief particularly in terms of time and cost.

The assessor may want to examine examples of document control methods for other individual projects. From the point of view of the project quality system the documentation should include actions to be taken to check that what is required has been done correctly. A document control system requires the establishment of a method of recording the outward and inward transmission of important information including the acknowledgement of its receipt, and of registers to monitor these. Often these do not exist at all on a project basis, or even overall within even quite large firms, and notwithstanding the requirements of ISO 9000, which cannot be expected to apply to a project-specific situation.

The sub-contractor will be expected to have developed appropriate methods of cost control and 'reporting back'. The auditor must be satisfied that any cost limitations are fully understood and that methods are in place for dealing with any deviations and for corrective action.

Generally all the information required will be gained on this one visit. Copies of all relevant printed forms, procedure notes or manuals should be looked at and if possible will be taken away for reference in the report to the client.

Where manufacturing is involved as part of the company's expertise and project requirement it will also be necessary to visit the manufacturing premises and undertake an audit on similar lines to the foregoing, but in this case the twenty sections of ISO 9000 Part 1 or Part 2 and any third-party certificate will be relevant.

Discussion will be necessary to establish what procedures the company has to maintain quality records and any procedures that exist for improving standards and reporting back to senior management – if any. It is particularly important that any such procedures are able to be translated from and into the project and can operate effectively on the site.

The conclusion of an initial audit will be a report to the commissioning client with recommendations in line with client requirements which may be:

(*a*) to establish the company's suitability to become a stakeholding member of a 'design team' for the specific project
(*b*) to establish the company's suitability to act as a sub-contractor within the partnership for the specific project either prior or subsequent to tendering for the work.

(In the latter case this may be as a nominated or domestic sub-contractor or as a sub-contractor for a parcel of work under a management contract and in either case with or without design or manufacturing implications.)

Thus, an opinion of the likely receptivity of the company audited to a 'partnering' approach will be assessed, along with the technical and logistic requirements of the project, and the sub-contractor should be assured that he or she will be invited to, and encouraged to, attend the initial partnering workshop when it is arranged.

5 Cultural changes for successful partnering

Culture: A social and intellectual formation – the totality of socially transmitted behaviour patterns, arts, beliefs, institutions and all other products of human work and thought – characteristic of a (particular) community.

Readers Digest Universal Dictionary 1992

A culture which leaves unsatisfied and drives to rebelliousness so large a numbers of its members neither has a prospect of continued existence nor deserves it.

Sigmund Freud – The future of an illusion

The underlying behavioural and motivational issues that are present in all projects and in most situations where human beings work are examined against the work of behavioural scientists Maslow, McGregor and others, and the background of construction conflict.

The recent findings of research carried out at Exeter University by Cartwright that cultural factors are developed either by design or default and can be measured are set in a project partnering perspective as a guide to establishing better project structures.

We have seen that philosophical and cultural changes are implicit in bringing about TQM in organizations, and how they are the most difficult changes that have to be made. Without these changes it is unlikely that the Kaizan approach to continuous improvement will develop or, indeed, that any real improvement other than on a very short-term basis will be sustained.

But what of the changes necessary to establish the partnering concept on individual projects? Will these be even more difficult? Easier? Impossible? Or will the approach within a company be easily transported into any given project?

To resolve these questions there are two opposing factors which we must consider for the construction industry.

Conflict in construction cultures

First, the global, or macro, state of the industry overall, and its individual components. Currently, cynicism goes from top to bottom, with the most extreme views gathering force as the idea flows from the designers through the main contractor to sub-contractors, and to sub-sub-contractors, where it consolidates as a hard cynical reserve – sometimes rock hard. This is the cynicism which causes vastly inflated claims on the basis that: 'Well, you never get the last £50,000 of your rightful claim anyway!'. Here there is a strong resistance to any change because of the disbelief that those earlier in the process will really change anyway. This starts with the collective view that from the client downwards everyone is greedy and wants more than they will pay for, and then seek to get it by crafty means within a contract created to catch out the unwary – and therefore generally the smaller – firm later in the process. Thus the hostile attitude (from past experience) is firmly embedded before any project starts.

Litigious culture

Within this macro analysis of a confrontational, litigious culture in the industry as a whole, each firm within the fragmented whole develops its own corporate culture developed from its own particular experience, its own role, and the style of its owners or managers. Frequently this develops a common attitude towards another group so that, for example, the name 'architect' invokes a particular response (and a different one) from clients, quantity surveyors, contractors and sub-contractors. Similarly, other groups are all expected to be tarred with the same brush and group attitudes exist towards and between all the other disciplines. Each firm within each group develops its own culture, then builds a hard shell around its core beliefs – right or wrong – and defends them against attack from the others.

For culture is about shared assumptions, beliefs, values and norms, and these drive patterns of behaviour that are and arise from 'the way we do things here'. 'Here' might be within a particular company, or in a sector, division or region of larger organizations, again depending on the personal influence of its current leader, manager or director, and in turn their exposure to the larger (or older) unit's culture.

Professional cultures

Within long-established professional practices – solicitors, surveyors, engineers – the partners will have stayed in and have progressed through the firm, because they have found the atmosphere to their liking. If not, the ambitious who feel they don't fit, or cannot wait, will have very quickly moved to a more amenable climate elsewhere, until, by the time they achieve partner status they have become part of and so consolidated the firm's culture, which thus becomes more established, more conservative, and less prone to change. For these reasons the technological aspects of progress within the industry have often not brought about the operational changes – or process changes – that are almost implicit in manufacturing and process industries.

Project culture

The second, and converse factor, that each new project brings is the potential that a group of firms, people, ideas and concepts have to be brought into existence – *de novo* – and thus the opportunity, hope and optimism that this one could be different, exciting, satisfactory and profitable could exist and be cultivated. Some may think that this is the optimism that weekly tempts gamblers into the National Lottery, or backing the 500–1 hot tip for the 2 o'clock at Ascot, but given the right leadership at the commencement of the project the teamwork necessary for success as a combined operation could develop. Teamwork such as that which produced success against all the odds on fortress Europe in 1944 and in the Falkland Islands in 1992. This success was repeated there later in 1993, in the construction operations to house the vastly increased garrison.

Project construction, much more a 'people' business, being both innovative in its products – every job has always been different – and yet conservative in its culture through the independence of its main component firms, resists and has been slow to change its management style. Thus, the task in many ways is no easier – unless designed, demanded and demonstrated from the top and from the beginning by the client. Culture change requires that the change in management style and leadership comes from the top.

Construction and its projects must have a consciously built people-structure and relationships to work at all, but to overcome the cynicism of its component firms and their confrontational culture a fundamental cultural change is required.

Culture change creates culture shock

Now cultural change is a process of reorganization in:

- beliefs and values
- perceptions and attitudes
- behaviour and relationships.

These changes will cause anxiety, can create conflict, but could provide opportunity on a project basis, one project at a time, through partnering. Partnering can produce a substantial improvement in performance and could easily reduce the overall costs by more than 30% through reduction in waste and other poor quality costs.

But it must be recognized that in construction as in other industries cultural change equals culture shock. When a cultural shock is delivered it often produces immediate reactions that:

- external changes are easier
- internal changes are harder.

But there is no doubt that new values must be accepted before old values are abandoned. In this respect, increased trust and responsibility could promote self-confidence and therefore self-reliability in all and particularly the later people in the process – the sub-contractor and the sub-sub-contractor. For, has not pride in workmanship always been a strong motivation in the traditional craft trades? Why should this be so?

Let us examine the views of some behavioural analysts to see whether we can develop a better understanding of the human factors behind confrontation, dispute and conflict in a management environment that could then be related to projects to develop a better project culture through its more successful management.

The effective management of projects requires effective direction of men and materials to achieve the particular objective – the client's building – but it is at the same time a process of 'getting work done through people'.

This is a definition given to me over lunch in the mid-1960s by Dr J. F. Dempsey who was then the Chairman of the Irish Management Institute. In 1936 he was one of the founders of the first Irish aeroplane company which he later developed as Aer Lingus and of which he was then its Chief Executive. This was a position he had held for most of the time since first becoming involved when the company – with a single De Havilland Dragon aircraft – was carrying six passengers on a daily trip of 200 miles between Dublin and Bristol.

Those who have observed the promotion of Aer Lingus will know that it described itself as 'the friendly airline', and those who have flown Aer Lingus will know that whatever other attributes it had, the airline staff, now 5000 strong and flying many millions of miles every year on daily schedules to every continent and most developed countries of the world, certainly reflect that adage. Jack Dempsey's attitudes, personality and influence were clearly reflected in his style of management of Aer Lingus.

Now the law itself, and certainly the common law and the many standard forms of contract developed for the construction process which reflect the customs and practices and established ways of the community in which it is based, could also be considered 'a social process'. But established ways reflect and encourage the conservatism of the British, and progress, or evolution, in the law in turn reflects the slow way in which social change develops.

In 'the direction of men and materials to a given end', we can recognize more easily the background implications of technique or scientific method which springs from the origins of modern management through 'work study' as developed by Taylor, Gilbreth and others. They realized that the development of techniques to improve performance required first the development of appropriate measuring techniques. Techniques to measure input against output, and the balance of requirements against resources, so that productivity can be established and better ways found to improve performance, without additional cost, within an organization.

These techniques can also establish benchmarks for later measurement of performance on projects.

But, finer measurement of performance in both physical and people systems is necessary and new terms and concepts need to be identified to be able to communicate the results and develop an understanding of them to other people, and across the disciplines involved in the project.

Work study produced terms such as 'Therblig' (Gilbreth backwards), work measurement, standard performance from standard rating, relaxation allowances, predetermined motion time systems, activity sampling, critical path networking, etc. and these reflect that management systems are technical systems for the direction not only of people but also of materials towards the stated objective. In the construction industry the application of materials also embraces the operation of machines and methods, which all have financial implications.

In the past the direction of men and women by financial incentives may have been sufficient, but in today's social structure motivation involves far more than just money. Today a literate and highly numerate community has greater freedom of choice, as a result of economic progress. Motivation means more than just money for workers in the latter part of the twentieth century.

So, today, the cultures developed and the conflicts they engender in construction must be recognized and somehow dealt with.

Analysis of cultures

Recent research by Jeff Cartwright and the Economic Psychology Research Group at Exeter University found: '. . . that it is the cultural work environment of an organization that determines the motivation of its employees . . . . And the psychological structure of a work environment consists of nine key motivating factors that are common to all the organizations studied'.

Their study covered over 100 interviews with chief executives, quality managers, personnel directors, shop floor workers and leading academics in the field of economic psychology at some of the top UK-based British, US and Japanese companies and revealed how inadequate ISO 9000 quality systems have been transformed by the adoption of total quality motivational methods.

Cartwright's conclusions were that external changes are always seen as easier whilst internal changes are harder – it is acceptable for the other fellow but not for me – I don't need to change. And within an organization the changes to process, method, and custom which have become part of the internal culture, are resisted.

For projects there is an opportunity to get the culture right first time as there are initially no particular methods or customs on the new site. The cultural style of the project leader and the way he or she approaches the management of the project is therefore critical and presents a golden opportunity for a successful 'new' culture on that site.

The Cartwright analysis found that a culture consists of three interactive systems:

- systems of actions and controls
- systems of knowledge and understanding
- systems of beliefs and values.

This drew on the work of Beattie in 1964, which found that when these three systems are in unison they gave meaning, purpose and direction to both life and work.

Conversely, industrial disunity is the core of social conflict (see Fig. 13). When alienated, sub-cultures will form by default. In construction the tradition has been to preserve these cultures for conflict and for battle in the courts or arbitration many months or years later.

The Cartwright psychological analysis of a total quality culture revealed that a good management controlled environment consisted of:

- professional management systems
- a belief in quality values
- a philosophy of continuous improvement in the pursuit of excellence.

This results in a particular cultural management style. Cartwright, like Hofstede in 1980 and 1991, found also that there were five variations in

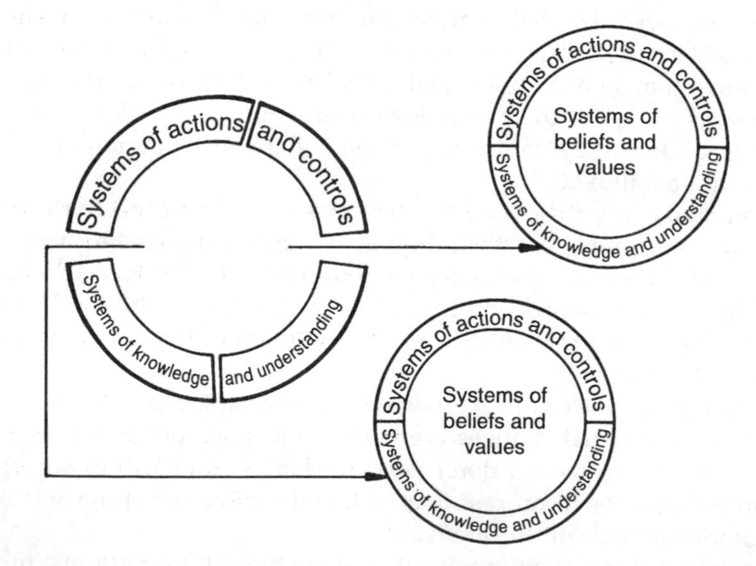

Figure 13. Cultural disunity is the cause of social conflict

cultures which came about from regional or organizational factors (in construction, probably also in professional disciplines and the nature of the different work done). These were:

- a style of leadership and power – whether autocratic, remote, participative, etc.
- an attitude towards gender, e.g. male–female: and their conventional roles/status
- collective or individual group cultures flowing from the roles and status which develop
- an attitude to change. Impervious (in long-established cultures) or adaptive and progressive as teams on construction projects must necessarily be
- the constituent factors and cultures such as a cosmopolitan talent by intelligent reaction.

Clearly it is necessary for the project directors and their manager to have cultural awareness when designing the management structure and approach to the specific project.

Cartwright identified seven cultural features within a 'standing' (as opposed to a dynamic project) organization that help to identify its culture. These were:

- distinctive to the culture
- satisfying to the participants
- productive to the participants within the culture
- inclusive (or exclusive) factors that help to develop the culture
- objective
- instructive of the participants
- continuous.

He also identified (like Deming before him) 14 commodities that were present in all world-class total quality organizations which were included in the study. These were:

- a strong corporate identity
- clear corporate objectives
- corporate ethical values
- market driven approach
- high quality control systems
- continuous improvement
- a distinctive management style
- health and safety hygiene
- teamwork
- unity of purpose
- freedom of information
- rational approach to problems

- positive workforce response
- corporate success in their chosen field.

Further analysis by Cartwright produced nine psychological characteristics that defined the organization's culture. These are:

- identification
- internalization
- instrumentality
- consensus
- rationality
- development
- group dynamics
- equity
- equality.

He further suggested that these motivating factors defined the negative and positive aspects of the organization's culture. The negative aspects arrived at by default whilst the positive aspects were created by design and therefore produced a management controlled environment that reflected management philosophy, beliefs and systems.

Cartwright then went on to apply the nine factor format which he says can produce a psychological profile of a work culture and, as in Figs 14 and 15, enables benchmarks to be created and later, by measurement of these factors, produce a further profile of the progress (or otherwise) of a particular culture. Thus, if we learn from Cartwright's work and apply its results to develop a project's total quality culture we must:

- assess (and measure) existing cultures in the organizations who will be retained to create the project
- identify the changes needed to make them sensitive to the proposed project culture
- design a partnering structure around the positive aspects of the contributing organizations' existing cultures
- introduce and train all leaders in the quality values and benefits – a highly important element of the initial partnering workshop

and then:

- monitor the situation at each stage and as we assess the progress reflect on the contributors to the culture of the project.

Maslow and motivation

Let us now look through the eyes of Maslow at the human element which is present in construction management situations: Dr Dempsey's 'getting work done through people'. What motivates people? What motivates designers in drawing offices? What motivates workers on a

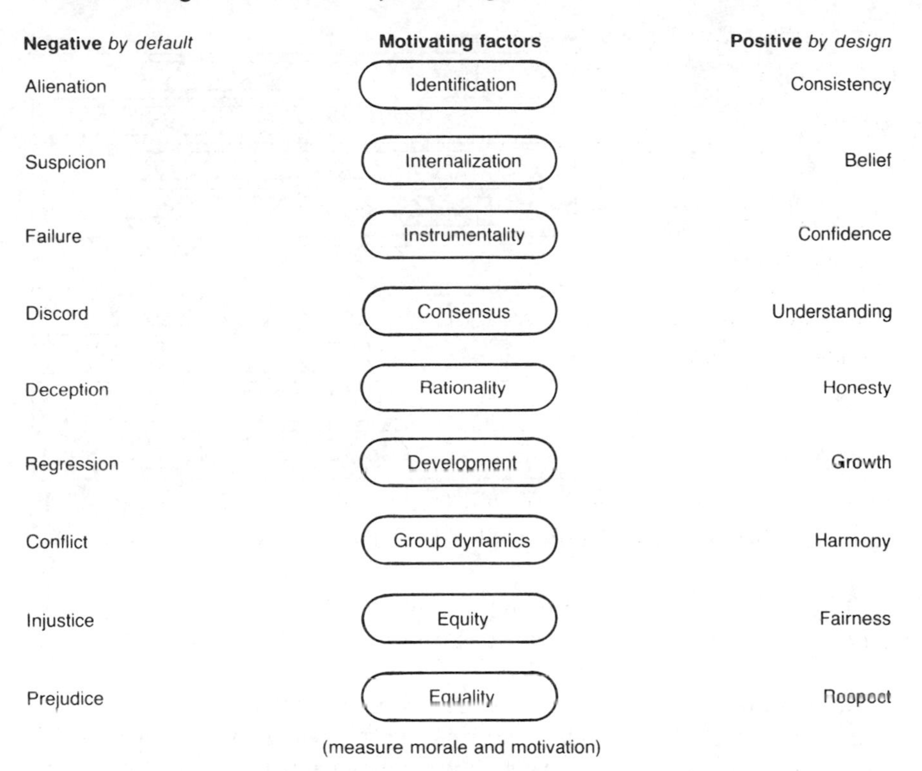

Figure 14. Cartwright's nine key motivating factors

construction site? This will depend partly upon the conditions under which they work and there will certainly be different attitudes to working in exposed conditions in the summer with the temperature at 30°C, to that in a similar situation in the winter at –10°C . . . or will there? As far as the work is concerned there may be no difference! In both it could be to get the work done and 'get the hell out of it' – or to reverse the order – until the temperature changes.

But the reaction of human beings, according to Maslow, depends fundamentally on the extent of their ascent up their basic pyramid (see Fig. 16).

Behaviour and motivation

Behaviour is the individual's total response to *all* motivating forces. One of which is the particular situation at a particular time, but Maslow suggests that all basic human *needs* can be expressed in a hierarchy of prepotency. That is to say, the appearance of one (superior) need usually rests upon the prior satisfaction of a lower or subordinate human need. The five levels in this hierarchy are, in ascending order, physiological, safety and comfort, social, egoistic, and self-realization.

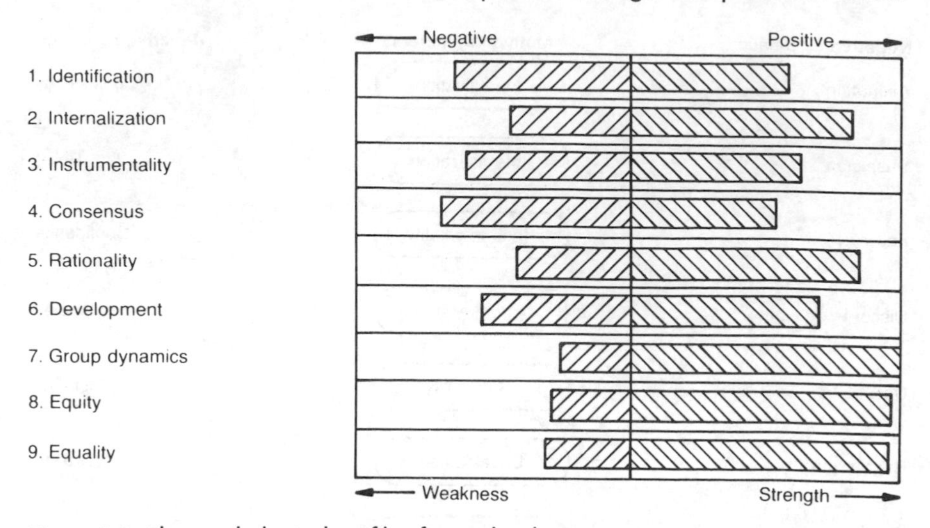

Figure 15. The psychological profile of a work culture

The theory postulates that man's animal wants are perpetual, and each drive is related to the state of satisfaction or dissatisfaction with the other drives.

Motivation is, however, human rather than animal-centred and is goal-orientated (in construction even more so) rather than 'drive'-orientated. All rational human behaviour is caused: we behave as we do because we are responding to forces that have the power to prompt – motivate – us to some manner or form of action. In a sense, therefore, behaviour *per se* can be considered to be an end result – a response to basic forces.

However, behaviour is actually only an intermediate step in a chain of events. Motivating forces lead to some manner or form of behaviour and that behaviour must be directed towards some end, that is to say, there must be some reason why we are responding to the motivating force. And that reason could only be to satisfy the force which in the first place motivated us to behave. Consequently, all human beings, whether they do so rationally or irrationally, consciously or subconsciously, behave as they do to satisfy various motivating forces, says Maslow.

The forces that motivate people are legion and vary in degree, not only from individual to individual but also from time to time. They range from ethereal and psychological to concrete physical, instinctive, and basic physiological forces, such as hunger, thirst and avoidance of pain.

Motivation is not synonymous with behaviour. Motivations are only one class of determinants of behaviour. Whilst behaviour is almost always motivated, it is also almost always biologically, culturally and situationally determined as well. We are, in short, the product of our environment.

Undoubtedly, physiological needs are the most prepotent of all needs. In

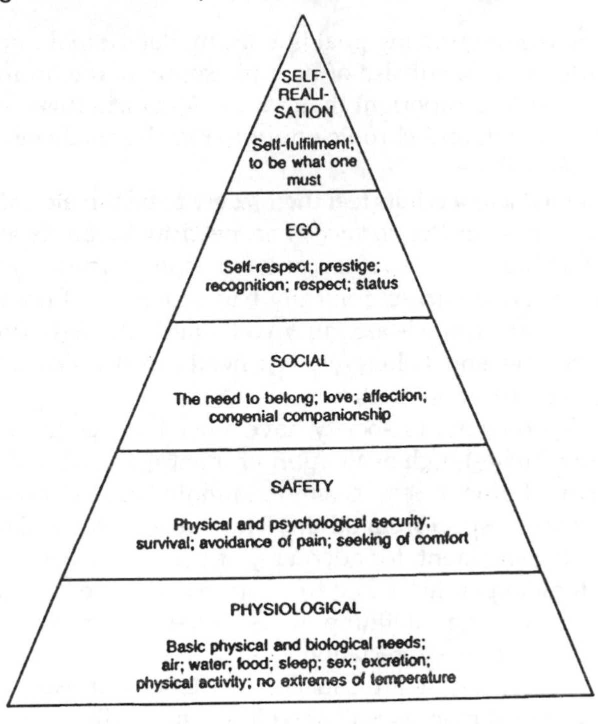

Figure 16. Maslow's hierarchy of human needs

the human being who is missing everything in life in an extreme fashion it is most likely that the major motivation would be the physiological needs rather than any others.

If all needs are unsatisfied the organism is then dominated by the physiological needs; all other needs may become simply non-existent or be pushed into the background, for consciousness is almost completely pre-empted by hunger.

Attempts to measure all of man's goals and desires by the behaviour displayed during extreme physiological deprivation are blind to many things. It is true that man lives by bread alone – when there is no bread. But, when there is plenty of bread at once other (and 'higher') needs emerge and these, rather than physiological hungers, dominate. When these in turn are satisfied, again new (and still 'higher') needs emerge, and so on. That is what is meant by saying that the basic human needs are organized into a hierarchy of relative prepotency.

Thus, gratification becomes as important a concept as deprivation in motivation. It releases the organism from the domination of relatively more physiological needs, permitting the emergence of other more social goals.

When physiological needs are relatively well gratified there emerges a new set of needs, categorized roughly as the safety needs. As in the hungry man, the dominating goal is a strong determinant not only of his current world outlook but also of his philosophy of the future. Practically everything looks less important than safety. A man, in this state, providing it is extreme enough and chronic enough, may be characterized as living almost for safety alone.

Unlike infants, when adults feel their safety to be threatened we may not be able to see this on the surface. The need for safety is an active and dominant mobilizer of resources only in emergencies, e.g. war, crime waves, neurosis, and such chronically bad situations. If both the physiological and the safety needs are fairly well gratified, there will emerge the love and affection and belongingness needs, and the whole cycle will repeat itself with this new centre.

All normal people in our society have a need or the desire for a stable, firmly-based (usually) high evaluation of themselves, for self-respect, and for the esteem of others. Self-esteem is soundly based upon real capacity, achievement and respect from others. These needs are, first, the desire for strength, for achievement, for adequacy, for confidence in the face of the world, and for independence and freedom. Second, we have the desire for reputation or prestige (defining it as respect or esteem from other people), recognition or appreciation.

Even if all these needs are satisfied, we may still expect that a new discontent and restlessness will develop, unless the individual is also doing what he or she is fitted for. A musician must make music, an artist paint, a poet write, if they are to be ultimately happy. What a man *can* be, he *must* be. This need we call self-actualization, the desire for self-fulfilment, to become everything that one is capable of becoming.

The specific form that these needs take will vary from person to person. In one individual it may take the form of the desire to be an ideal mother, in another it may be expressed athletically, or in painting pictures or in inventions. It is not necessarily a creative urge although in people who have any capacities for creation it will take this form.

The clear emergence of these needs rests upon prior satisfaction of the physiological, safety, love and esteem needs. People who are satisfied in these needs are basically satisfied people, and it is from these that we may expect the fullest (and healthiest) creativeness.

McGregor's theory X and theory Y

Douglas McGregor takes this behaviour pattern on into what he postulates as theory Y – that people are self-motivated and will respond to what Drucker called 'management by objectives' in contrast to 'management by control' (theory X) which he says results in people seeking to fulfil their social and self-fulfilment needs away from the job. Theory X, the more conventional management view, is that management is responsible for organizing all its resources for economic interests and this means that people are 'directed' to fit the needs of the organization. Without this firm direction people would be passive since they are by nature idle, lacking ambition and resistant to change.

Construction workers, perhaps because of their inherent job satisfaction – the carpenter who lavishes his skills on creating mouldings on doors, frames and staircases, the bricklayer whose prowess with an elaborate decorative brick bond – will be seen and admired by the generations that pass by his work. All appear to fit theory Y propositions better than they do theory X. Even more so do many think that every architect and most designers are all the time engaged upon fulfilling their Y needs at the top of the hierarchical pyramid!

But, because building construction (as illustrated in Fig. 3) presents a basic situation where the progress of any project involves many people whose objectives are widely divergent or perhaps convergent and on a collision course, the behaviour of the parties is often the result of these other factors present in conflict situations.

Cultural and group influences on behaviour

How an individual behaves will also depend upon his or her relationship with those other individuals with whom he or she is in daily contact – the direct and indirect cultural influences.

In the 19th century the freedom of choice for the average worker was very limited. Economic freedom existed for very few beyond the employer. In the early days of the Industrial Revolution in the 19th century individuals could be considered to be hemmed in by the organizations and people around them (Fig. 17)

In the 20th century economic conditions allow far greater freedom of action for the employed individual. The same circle of relationships between family, friends, church, trade unions, the employer and the state exist but their influence, either constraining or stimulating, has changed, at least as far as most individuals are concerned, from that experienced by a worker during the early Industrial Revolution (Fig. 18).

If we take this analysis one stage further into, say, a dispute situation in the 1990s, we find the position is different again. The relationships between the individual and the institutions, whether family, friends or authorities, are perhaps less immediate and influencing than they have ever been.

Since the Second World War children have been encouraged by their schools 'to do their own thing' and the economic climate of Britain in the 1990s allows them to exert their free will to choose, even if the choice has led to thousands of people spending thousands of pounds to become drug addicts, and so to be no longer master of their own destiny and in a horrifying way!

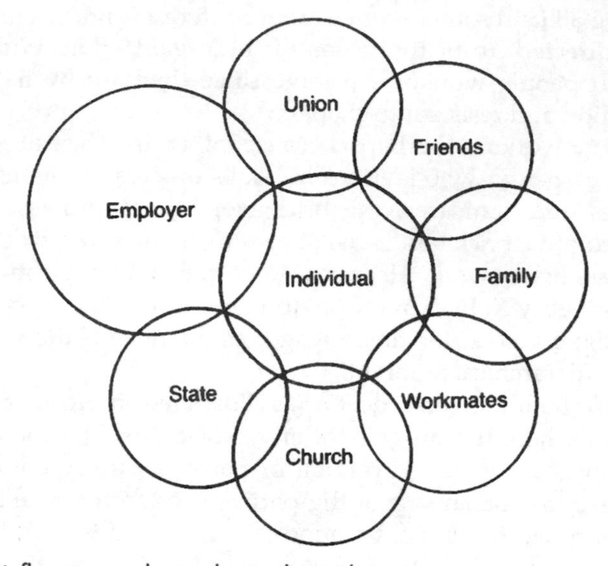

Figure 17. Influences on the worker in the 19th century

But the media generally, and TV in particular, are now a greater influence on the average person. There is a much more direct, if one-way, communication with events such as we witness daily on TV in the home. The selection of shots by the TV producers have a more direct influence than pamphlets or the written word. Seeing is, after all, believing, as the subtleties of selection, first by reporters and then by producers and the programme presenters, are not so apparent. But clearly they have an effect on the viewing individual's attitudes and even actions in conflict situations (Fig. 19).

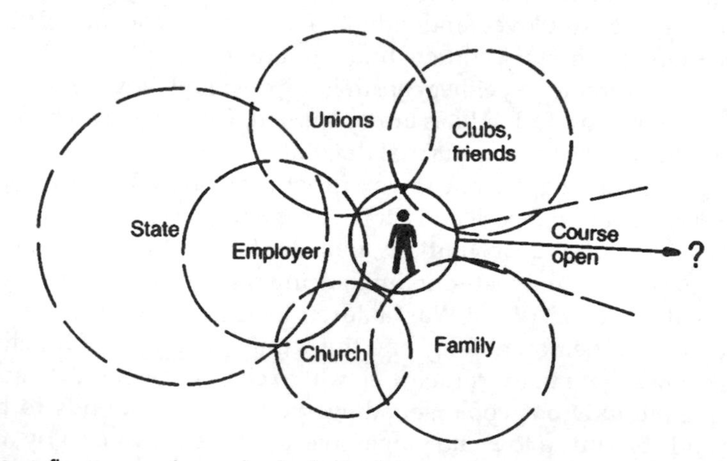

Figure 18. Influences on the worker in the 20th century

Let us take this analysis one stage further and consider the likely implications on the two parties to a construction dispute. We can then see the benefits of the introduction of a neutral adjudication tribunal to bring about a solution to the dispute, providing that it recognizes the management culture of construction and studies the contributory elements that caused the situation which may have flowed from materials, machines, methods or money, but takes account above all of the motivation of men, and the working cultures within which they have developed (Fig. 20). This topic and resolution techniques will be revisited in Chapter 7.

Construction projects inevitably involve conflict situations. Problem situations arise from technical, climatic and logistic events, but disputes are caused by people and the motivational, behavioural and cultural implications on their actions will play an important, if not always visible, part in their creation – and so of their resolution.

The concept and practice of project partnering has been developed to create on a project-by-project basis, a culture which reinforces positive motivation at a high level in the Maslow hierarchy, recognizes the relevance of McGregor's theory Y and establishes basic benchmarks for measuring and developing a positive design culture from Cartwright's findings. Putting the final element in place in the holistic approach of TQM, for:

- procedures – will not work without knowledge of:
- practices – which must reflect an overall effective:
- process – which in turn is based on good:
- principles – which flow from:
- the philosophy and ethos of:
- partnering.

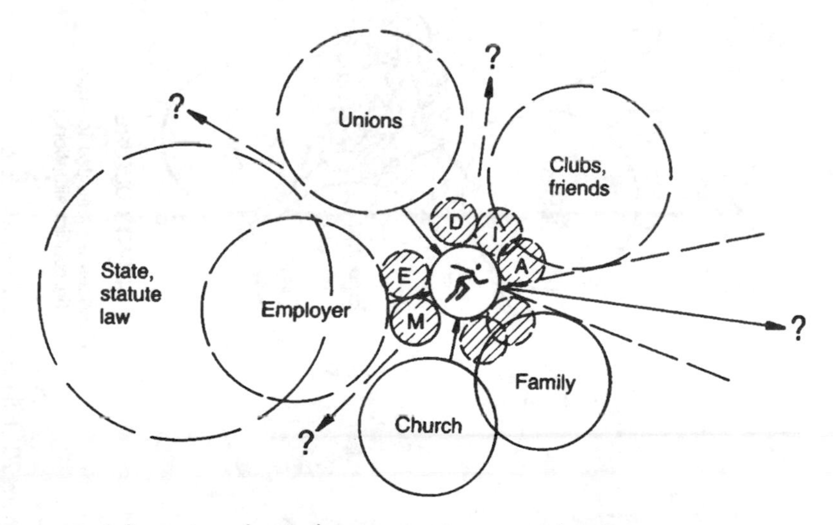

Figure 19. Influences and attitudes to action in a stress situation

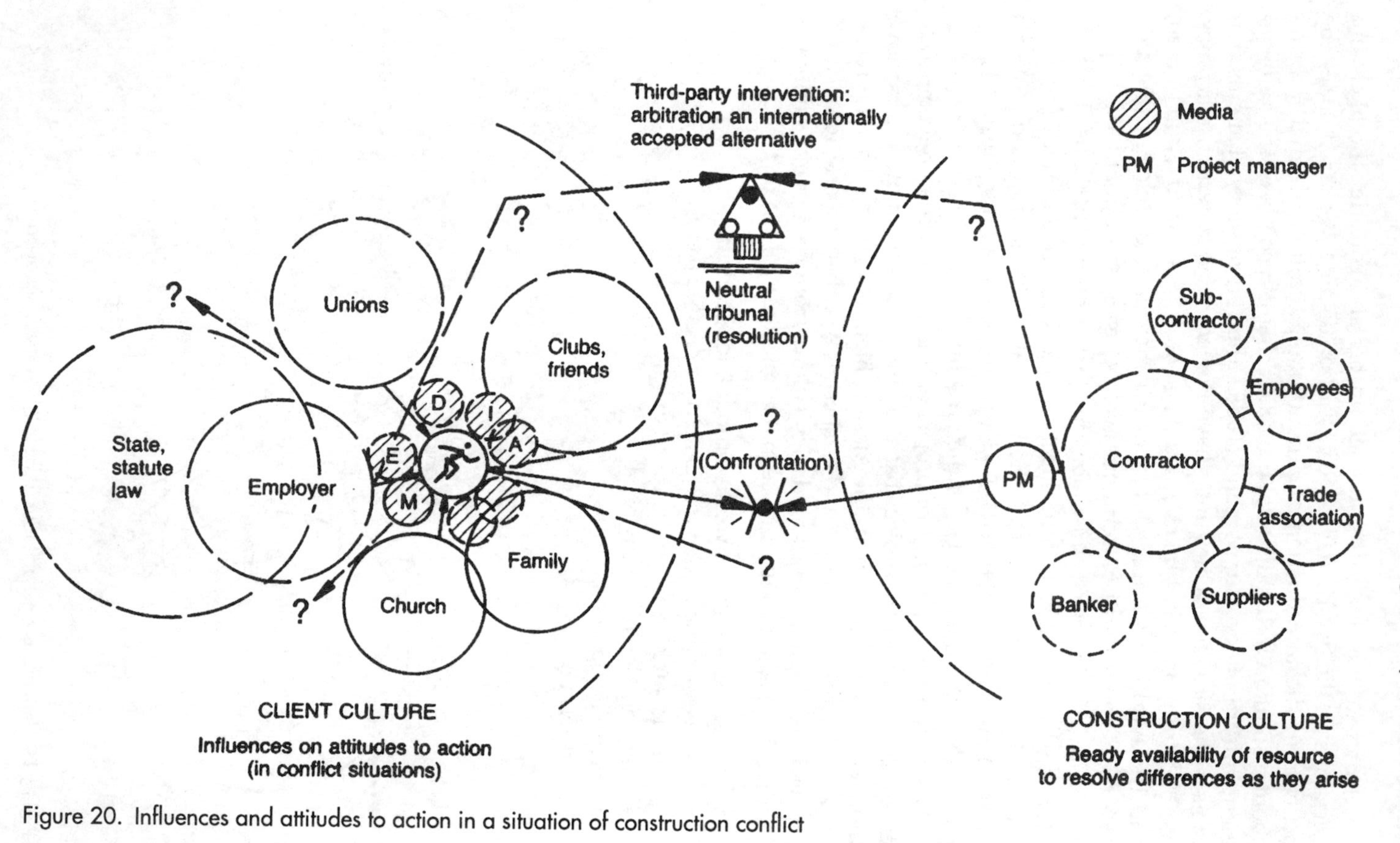

Figure 20. Influences and attitudes to action in a situation of construction conflict

Structuring the project

The fundamental premise of a revolution is that the existing social structure has become incapable of solving the urgent problems of development...

Leon Trotsky

This chapter attempts to fill out three areas and techniques of project management essential to the success of the partnering approach. First, developing a project structure to better define the roles and relationships of the many firms and individuals who will be involved in the development of the project from conception to completion. Then, it looks into the problems of effecting real communication to encourage commitment from conviction, and sustaining the message with an analysis of the nature of communication through meetings – as methods of group transaction, and finally it stresses the practical difficulty of persuading the 'last in line' – the specialist sub-contractor – to believe in the concept and join the team.

Now that we have looked at the philosophy that lies behind partnering and considered the principles of quality management and their suitability, sometimes after adaptation to the construction project, we must now begin to apply these to the project, its organization structure and the process by which human beings will bring about its creation.

The amateur client, coming new to this project process, cannot be expected to know all about the process that he or she may only be involved in once, and only in recent years has project management been developed and studied by some of the other factions involved for building projects. In building projects TQM is synonymous with all the concepts and precepts of good project management. Those attitudes and techniques used to satisfy the client's requirements for quality in function, aesthetic, cost and time. They must be used to establish and develop the project culture.

Organization and responsibilities in construction projects

Project quality will only result from good project management. An examination of methods, techniques and activities shows that poor productivity in the industry results more frequently from a failure in human relations and proper organizational descriptions rather than from technical inadequacies. This problem is often aggravated by the form, timing and nature of the contracts between the parties and their partial ignorance or understanding of their intentions and legal meaning.

The last 15 years has seen many technical changes in building construction, components and design. These changes have not been adequately recognized or reflected in the organization of either the professional or commercial sectors of the industry.

It is difficult for firms traditionally engaged in the industry to look at the problem overall, since few have the opportunity of managing the project overall. This problem has led to the growth of the all-in service and package deal or 'turnkey' contractor, considered by many contractors, as well as professional firms, not to be in the best interests of the building client. It represents the evolution of one new form of organization to cope with the overall problem of productivity and the giving of an efficient service to its clients by the industry. The form of service it offers is not necessarily any more efficient than a properly considered organization along the lines of the traditional relationships. The fact that many building projects must inevitably continue to be 'one-off', or as a civil engineer said recently: 'most projects in civil engineering and building are development projects', and by their very nature make the techniques of industrialization more difficult to apply.

The sequence of events

First there is the need for the client, that is the person or group requiring the building, to define what are the requirements for the building in the form of a 'brief'. This has to state what is to be done, first in broad functional terms and then in more detail.

Secondly, in the sketch or outline design stage a hypothesis is developed as to the possible building solutions to this problem. This in turn is subjected to a more detailed examination to see that all the various parts of the building fit snugly into the plan – the detail design stage; then, still later, comes the 'instructions for production' – the production or working drawings. These are followed by an examination and grouping of the resources which will be required to meet the plan – the skills of quantity surveying, and then by the contractor in planning and critical path analysis; this then becomes the final programme for the execution of the project.

Another way of looking at this problem is the 'who' are involved in the development of the building project. Here, in the planning stage, is a mammoth task of communications and coordination. Someone has to see that the requirements of all these people involved in this stage will be met by the building design. Some of those involved have statutory functions and the power to exercise even more control over the building than the client. The problem then has the added requirement of persuasion as well as communication!

Thus, we have an analysis of the technical problems involved in a project. Much less consideration has been given to the organizational implications and of the contract methods necessary to give effect to these methods. Even in companies where conscientious attention has been given to the company organization and structure, less thought has been given to the organization of 'the project': yet this is the raison d'être for every firm in the industry.

The differences between project management and the static management of an organization would seem to be the difference between the static traditional organization of military arms, wings, fleets, commands, divisions, brigades, regiments, companies, platoons and sections with the 'dynamic' organization of units brought together for a combined operation, where soldiers rub shoulders with airmen who provide both the air cover and the transport to the operation, and later with the sailors who have provided support and perhaps delivery to and evacuation from the operation.

In this 'combined operation' the objective was clearly defined as were the roles of all the participating units each with their own sub-objectives. The operation worked because the structure for the operation was especially laid down and drilled into the participants so that it was understood by all.

Confusion over objectives

In a building operation those engaged are frequently left to work out their own objectives – in the light of previous experience, often at a very low level. Since these experiences will have been different it is highly likely that their procedures will differ and confusion over objectives will arise. Further complications arise, for the companies and even the people involved in the combined operation of a building project are likely to be involved on a number of different projects at one and the same time. This diversification may even apply in the professional offices, and it is by no means unusual for the architect, surveyor and structural engineer, as well as the contract managers for the various sub-contractors, to have five or six, or even more, projects under their control at the same time, and it is quite likely that some of these projects will be the subject of different forms of contract and certainly of different patterns of management and project organization.

That various arrangements are commonplace underlines the importance of a specific organization pattern for each project. It will be seen that in all cases there is a dependence of the client upon the building engineer and designer (the architect), the structural and mechanical engineers and

Figure 21. All pulling together

the quantity surveyor, and they in turn must rely upon the general contractor and the various sub-contractors as well as suppliers and manufacturers. In fact, complete *interdependence* of the whole building team is essential if the client's needs are to be economically satisfied.

There is the need to ensure a balanced progress towards a stated objective if the project is to succeed. This is the first management task – to set up the (quality) plan for achieving the objective. This plan must itself be balanced; a balance of function, aesthetic, cost and time, all of which are present in some degree on every project. Whoever is the 'first man in' must set up the strategic organization for the achievement of these objectives by the other professionals involved, once they have been defined by the client and interpreted, through the brief, into at least a schematic design, which will determine the broad nature of the construction task – long-low, loose fit, or a monolithic concrete structure rising perhaps 450 feet in the air in a point or slab block.

Many of these problems involve human relationships and produce conflicting loyalties. Within such a situation the organization pattern often cannot be an 'ideal' unified, management structure, because the contract between building owner and contractor has not yet been formed whilst the separate contracts between the building owner and the individual members of the design team may already have been established.

In this way the players are often hampered by hierarchy, departmental, disciplinary and cultural boundaries, political issues and a willingness to settle for curing symptoms and winning the claim (Fig. 21).

Some means of unifying the objectives and clarifying attitudes towards a project are clearly necessary. The problems should be considered afresh for each type of project, and partnering arrangement (Fig. 22).

Project organization – workable management structure

The efficiency of any project or organization must start with top management and in the building industry this means with the client. Clients will generally be inexperienced and no matter how efficient the remainder of the building team may become, unless the client can be taken along with the project its later management can be difficult and sometimes impossible.

The traditional organization of the building industry has always had regard to this inexperience of the client; indeed, the pattern of the industry and the forms of contract which have grown up have had this at least as a background raison d'être.

There are several basic reasons why building projects will continue to be 'one-off'. Requirements will often be specific to an individual client; the site will be an individual one and final assembly must always be carried out on

Figure 22. Cooperation is working in unison

that site. For these projects the 'typical' or traditional organizational system has evolved. But many structures and systems are possible and some will be more suited to the conditions of a particular project than others. What is important is that conscious attention must be paid to the organizational relationships of every key player in the whole construction team.

'Organograms' should be produced, responsibilities defined, and contractual relationships and communication channels made clear, and preferably unified – for crossed lines of communication contract conflicts make.

To illustrate a solution to a typical one-off situation, the author is indebted to all those involved in the City of Plymouth's Civic Theatre. Figures 23–29 are taken from The Theatre Royal, Plymouth, a project carried out successfully in 1979–1981 and one providing much pleasure to the client and the community since its completion in 1982.* In this case 'the community' includes the civic corporation, the ratepayers, the actors, and the staff of the theatre, and even theatre critics – all of whom have written in glowing terms of the end results as arguably the finest post-war theatre in the UK. During its creation too, the theatre reflected good project management, carried out in the framework of good human relations, before any formal approach to partnering had been developed.

Were these mandates to be drafted today within the partnering concept each would carry reference to the approach, the charter, the procedures and everyone's part in them. These would be expected to feature in the manual contents list and in all parts of the manual, for example:

Part 1 Introduction
Part 2 In the full team meetings and the contractor's site meeting
Part 3 All mandates for the stakeholders
Part 4 Adding partnering assessment, reporting and action plans.

Communications within the project team

... evil communications corrupt good manners.

Corinthians 15.33

Having established the organization structure and the relationships of those involved it will be essential to set up an effective communication system. A great deal is talked about 'poor communication' and lack of

*Some of these illustrations and more of which feature as a case study on pages 104–130 of *Managing Construction Conflict*.

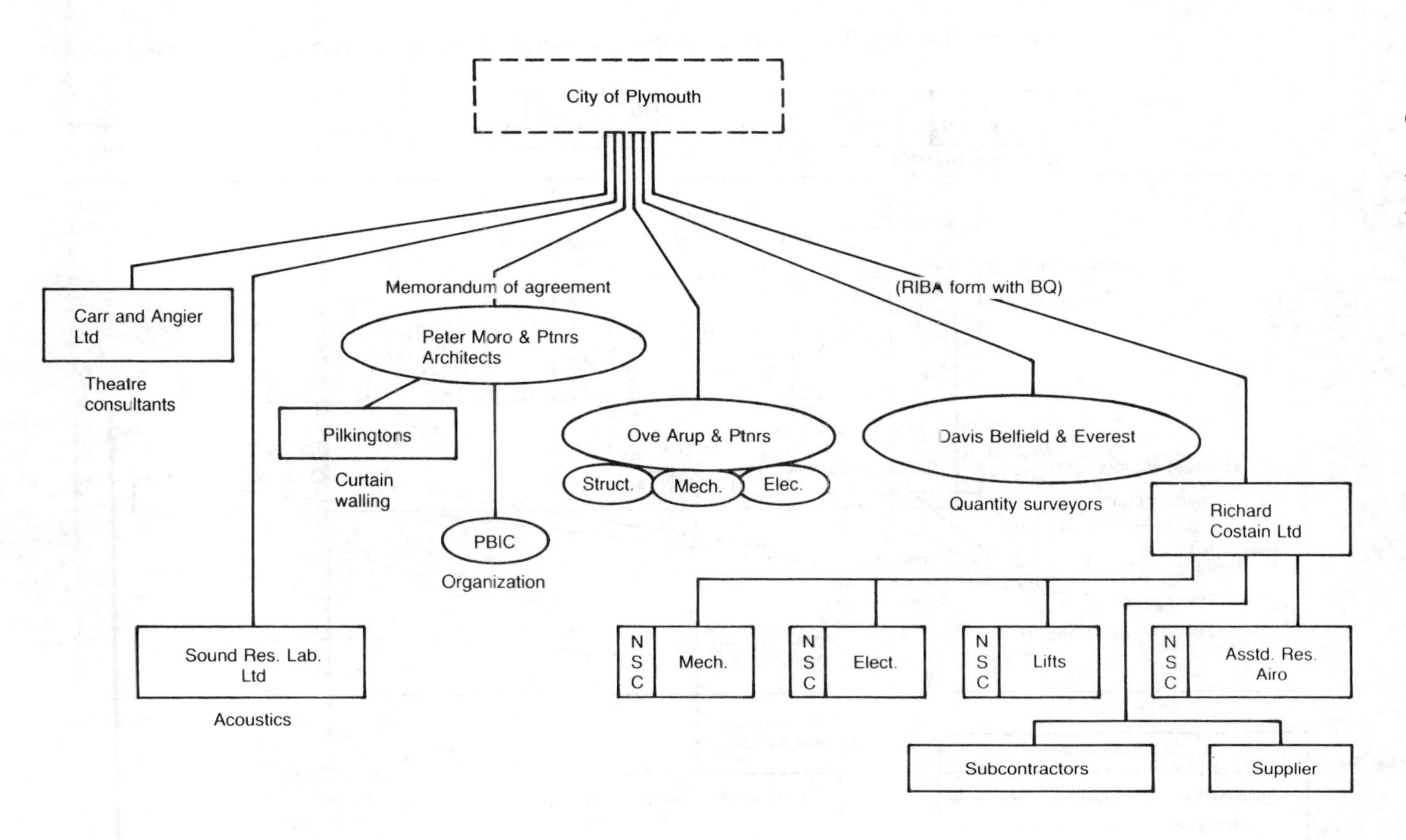

Figure 23. Plymouth Civic Theatre – nexus of contracts

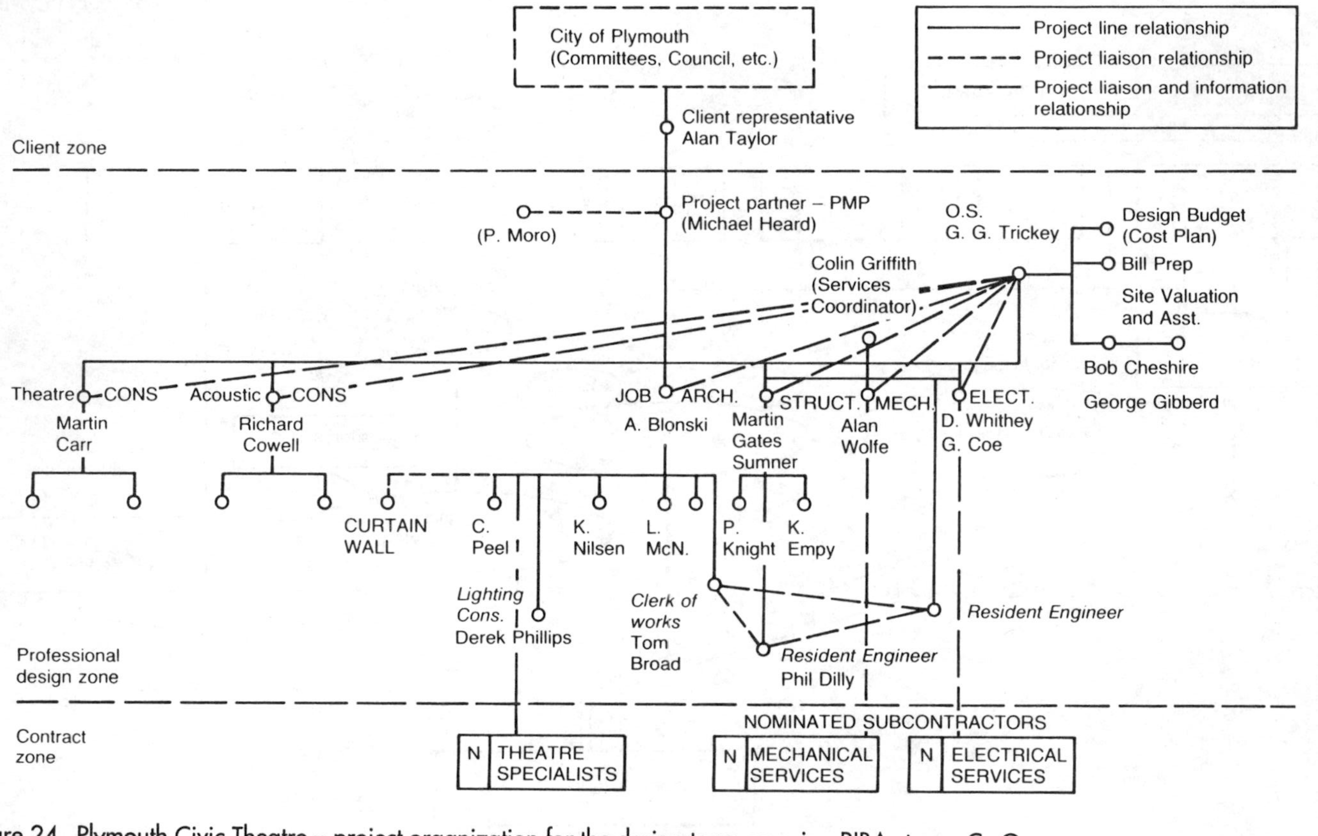

Figure 24. Plymouth Civic Theatre – project organization for the design team covering RIBA stages C–G

TYPICAL CIVIC THEATRE PROJECT OPERATIONS & PROCEDURES MANUAL

INDEX TO MANUAL

PART 1 - PROJECT ORGANISATION STRUCTURE AND RELATIONSHIPS

- INTRODUCTION
- DIAGRAMS - A) NEXUS OF CONTRACTS
  B) PROJECT ORGANISATION - RIBA STAGES A-B
  C) PROJECT ORGANISATION - RIBA STAGES C-G
  D) CONSTRUCTION ORGANISATION - RIBA STAGE H

PART 2 - FUNCTION OF MEETINGS DURING CONSTRUCTION PHASE

i) ARCHITECT'S SITE MEETING
ii) QUANTITY SURVEYOR'S SITE MEETING
iii) CONTRACTOR'S SITE MEETING
iv) DESIGN TEAM MEETING - FULL TEAM-ROUTINE
  - AD HOC

PART 3 - PROJECT MANDATES

	* CLIENT'S REPRESENTATIVE	-	(STATE NAME)
	* PROJECT MANAGER	-	(OF CURRENT)
	* JOB ARCHITECT	-	(PERSON)
	ARCHITECT'S DESIGN TEAM TEAM LEADER	-	( )
	CLERK OF WORKS	-	( )
CONSULTANTS	* STRUCTURES DESIGNER	-	( )
	MECHANICAL SERVICES DESIGNER	-	( )
	ELECTRICAL SERVICES DESIGNER	-	( )
	THEATRE MANAGEMENT (SERVICES) DESIGNER		( )
	ACOUSTIC SPECIALIST	-	( )
QUANTITY SURVEYOR	CONTRACT VALUATION SURVEYOR	-	( )
GENERAL CONTRACTOR	CONTRACT MANAGER	-	( )
	SITE MANAGER	-	( )
SUB-CONTRACTOR	CONTRACT MANAGER	-	( )

PART 4 *PAPERWORK AND FORMAL RECORDING PROCEDURES

i) MINUTES OF MEETINGS - DESIGN TEAM
  - ARCHITECT'S SITE
  - (NOTE: CONTRACTOR'S MEETINGS WILL BE DETERMINED AFTER CONTRACTOR APPOINTED)
ii) CLERK OF WORKS WEEKLY REPORT
iii) CLERK OF WORKS DIRECTIONS (RESIDENT ENGINEER'S DIRECTIONS)
iv) ARCHITECT'S INSTRUCTIONS
v) ADVICE OF CONTRACT VARIATION
vi) CONTRACTOR'S QUARTERLY REPORT (TO CLIENT DIRECT)

PART 5 PROJECT DIRECTORY - ADDRESSES/TELEPHONE NOS./ETC.

Figure 25. Contents list of loose-leaf project operations manual

PETER MORO & PARTNERS	OPERATION & PROCEDURE MANUAL	
PLYMOUTH CIVIC THEATRE ORGANISATION STRUCTURE	Ref:	POP/
	Orig:	July 1979
PROJECT ORGANISATION STRUCTURE & RELATIONSHIPS	Date Revised:	

1. The successful completion of the project, particularly in terms of time and cost, will depend upon the effectiveness of the organisation brought into being to coordinate and control the efforts of the whole Design and Build Team. Some of those involved will have no other work commitments during the operation, others will have several. All will have, however, permanent commitments and responsibilities to their 'parent' organisation and conflicts of responsibility or objective are therefore possible.
2. These project procedures are intended to define the management objectives, roles, responsibilities and relationships of all those involved in the project and so clarify communications and authorities for all involved.

   They are intended to clarify and not conflict with the contractual obligations entered into between the parties.
3. The assumption behind each 'job mandate' is that the person appointed to the task accepts the role, responsibilities and relationships shown in this organisational procedure, and has had the approval/confirmation of his 'parent' organisation to the mandate.
4. It is accepted that the individuals fulfilling some of the roles may change; for this reason names are given in (parenthesis). Changes in post, which should be notified in advance to the Job Architect, will not therefore alter the mandate. Where the incumbent of the post on the project structure draws his authority essentially from his own organisation it will be generally appreciated if any changes intended by that organisation are discussed, prior to their becoming effective, with the Architect's Project Partner and Job Architect.
5. On receipt of this Procedure Note everyone named should therefore consider the implications and if anything is either unclear or apparently unacceptable for some reason related to contractual or other obligations which are not known to the Architects, a draft of the amendment desired should be sent to the Job Architect within 7 days.
6. The effectiveness of the operation will result partly from efforts of the individuals involved, partly from the mechanism of their formal relationship and partly from the paperwork flow instructing and recording their activities. The procedures by which the project will be controlled are therefore grouped into:-

   A. The purpose, function and authority of the various individuals involved – Project Mandates

   B. The purpose, function and authority of meetings held – Meeting Mandates

   C. The paperwork flow and formal recording procedures.

Figure 26. Project structure and relationships

Figure 27. (*Right*) Project mandates – meetings (*continued overleaf*)

PETER MORO & PARTNERS
PLYMOUTH CIVIC THEATRE
ORGANISATION STRUCTURE
PROJECT MANDATES – MEETINGS

OPERATION & PROCEDURE MANUAL
Ref: POP/
Orig: July 1979
Date Revised:

1. During the progress of the work on site (through RIBA Plan of Work Stages H (I, J, K, L) progress and coordination will be assisted by the holding of formal meetings of different groups involved in the operations. These are:-
   – Architect's Site Meeting
   – Q.S. Site Meeting
   – Contractor's Site Meeting
   – Design Team Meeting

   Their purpose, status and functions are as follows.

2. Architect's Site Meeting
   i) Purpose – To review progress and so consider in advance problems likely to arise in the future which can be resolved by the presence of the responsible designers and the contractor.
   ii) Status –
      a) This is the principal meeting to effect coordination and implementation of the Design by the Contractor's Building Team. It will be held regularly, normally at monthly intervals, and at such other times as considered necessary by the Job Architect.
      b) It will be attended by every project engineer/consultant and by the Contract Valuation Surveyor, or by their representative who should normally be authorised to discharge the engineer's responsibilities, and by the Contractor's Site Manager and approved support staff, and at appropriate stages by the sub-contract managers for each nominated sub-contractor.
      c) The Job Architect will take the chair.
      d) The Client Representative, Architect's Project Partner, and a senior representative of consulting firms may attend in an ex-officio capacity.
      e) The Project Architect may invite (from time to time) other representatives from firms involved as he considers desirable.
   iii) Minutes will be taken and issued to all attending by the Architects.

3. Quantity Surveyor's Site Meeting
   i) Purpose – The Contract Valuation Surveyor will visit the site from time to time to carry out his duties. To assist him he may call either regularly or as he considers necessary a Site Meeting of the Contractor's Quantity Surveyors and those appointed by any nominated sub-contractor.
   ii) Status –
      a) These meetings will be concerned only with the valuation of the works and will be the sole concern of the Contract Valuation Surveyor.
      b) They will not normally be attended by other members of the Design Team, and will have no formal status under the building contract(s).
      c) The Valuation Surveyor may issue minutes should he choose to do so, in which case a copy will always be sent to the Job Architect.

4. Contractor's Site Meeting
   i) Purpose – The Site Manager will hold regular site meetings of his site team, nominated and other sub-contractors to effect coordination of the Construction Team.
   ii) Status – The meeting will be concerned only with matters of management which are the General Contractor's sole contractual responsibility. Any minutes that are taken will not be circulated to the Design Team.
      The General Contractor will decide who should attend and the Job Architect

or his representative, the Design Engineering Consultants or their representatives, the Quantity Surveyor's Contract Valuation Surveyor or his representative, will attend only at the express invitation of the Contractor's Site Manager.

5. Design Team Meeting
During the progress of works on site, particularly in the early period, there will be an overlap with the completion of the design. Meetings of the Design Team which have taken place throughout Phases D – G may therefore continue.
   i) Purpose – The purpose of these meetings, as was the case during the design phases, is to ensure continued coordination of the design process.
   ii) Status – The meeting will be attended by the nominated representative of each of the Design Consultants and the Quantity Surveyors, by the Job Architect, (who may deputise for the Project Partner in taking the Chair), and by the Architect's Team Leader, who may be accompanied by members of their respective design teams.
   Decisions taken, particularly as to functional needs, time and cost, will be minuted and circulated to all of the Design Team.

At later stages in the constructions of the works meetings of the Design Team may well be held separately from the Architect's Site Meetings to clarify points of design or prepare design modifications within the existing scope of the contract, or to deal with Variations. The Job Architect will normally take the Chair on these occasions.

6. Ad Hoc Meetings
Additional 'Design Team' Meetings between two or more, but not all members of the Design Team, may be held to consider specific problems that may not affect all. These will follow a similar format, be numbered in sequence and minutes will be issued to ALL who receive Design Team Meeting Minutes.

Figure 27. *(continued)*

communication in work situations but there is little in the way of guidance on how this can be improved – certainly in relation to the building project.

To help us understand the nature of communications in groups generally, and also in the groups of people in the building industry involved in its daily project work, we can approach the problem and analyse it from three starting points.

- The nature and function of the group.
- Members of the group.
- Behaviour in the group.

There are several other ways of dealing with a study of construction communications but perhaps this approach is the most helpful. An examination of these three aspects, from which various points arise, can help us to improve our project performance as a coordinated, motivated team. It is, of course, possible to analyse any sort of animal behaviour, including that of human animals, to a very high degree, but whether this alone improves us or teaches us to do better is a matter for conjecture – without effective leadership it may not.

PETER MORO & PARTNERS	OPERATION & PROCEDURE MANUAL
PLYMOUTH CIVIC THEATRE ORGANISATION STRUCTURE	Ref: POP/ Orig: July 1979
PROJECT MANDATE – CLIENT REPRESENTATIVE	Date Revised:

1. The Client shall appoint a Client Representative to act as the focal point for all communications between Client and Design Team on the one hand and the Construction Contract Team on the other.
2. All contractual matters shall be submitted by the Project Partner to the Client Representative and his authority will be sought on any matter connected with a variation to:-
   i) the Client's Brief;
   ii) the authorised design, or
   iii) amendments required as a result of constructional necessity or to the authorised programme for completion.
3. The Client's Representative may be invited to attend Architect's Site Meetings. He will receive copies of the minutes of the Architect's Site Meetings, Architects Instructions, Contractor's Quarterly Report and advice of contract variations.
4. He will be responsible for ensuring that payments are made by the Client to the General Contractor, to the Architects and to other members of the Design Team in accordance with their contracts with the Building Owner.

Figure 28. Project mandates – client representative

The nature and function of the group

When we find ourselves part of a working group or when we are considering setting one up, the first question we should ask ourselves is: 'what kind of group of people should this be?' There are a number of possibilities, amongst which are as follows.

A committee

In the building industry and its allied professions we do not often encounter this kind of group in everyday work. In our private lives, however, most of us have at one time or another been a member of a committee. Those committees that do exist in the industry are few, and many people think that they ought to be fewer. However, the function of a committee is simple and well known. It is a method of arriving at a major decision involving participation by interested parties, or their representatives, in the decision-making process. A participation which might also help later in the implementation of the outcome.

Although a very common method indeed, it is not always wisely used but, in its proper place and in the proper hands, it can be a very useful

PETER MORO & PARTNERS — OPERATION & PROCEDURE MANUAL

PLYMOUTH CIVIC THEATRE
PROJECT ORGANISATION

Ref: POP/
Orig: July 1979

PAPERWORK & FORMAL RECORDING PROCEDURES

Date Revised:

In addition to the Conditions of Engagement which define contractual obligations of the design team to the client and the general contract on the RIBA Form which is the basis of the contract between the Contractor and the Client, the project will be administered and recorded by minutes and paperwork control systems as follows:-

i) Minutes of Design Team Meetings.
ii) Minutes of Architect's Site Meetings.

both of which will note all items discussed and record formal decisions taken at the meetings.

iii) The Clerk of Works will issue Directions on the standard RIBA Form (attachment A), which will also incorporate Directions initiated by Resident Engineers for the Design Consultants which are authorised by him.
iv) The Clerk of Works Weekly Report Form (attachment B).
v) Architect's Instructions issued on the RIBA form, which will verify Clerk of Work's Directions and also indicate the authority or cause in order to assist with the determination of items which will constitute a justifiable Variation to the contract in either time or cost.
vi) Advice of Contract Variation, ACV, which will be issued by the Architect to inform the client of agreed contractual variations (this assumes that the client's authority will have been obtained earlier to the Variation if it is not the result of an essential requirement for constructional reasons).
vii) The Contractor's Quarterly Report, which will be issued direct by the Contractor to the Client, with a copy to the Architect.

The distribution of all paperwork should be shown on the document itself as indicated on this Procedure Note. The coding for distributions are:-

F	–	Full distribution;
ADT	–	All Design Team;
ASM	–	All attending Architect's Site Meeting;
CR, Th, Ac, S, M, E, Arch, QS, COW		Particular adressee only, or additional to other circulation, e.g.: ADT + COW = All Design Team plus Clerk of Works; or ADT – CR = All Design Team less Client Representative.

Figure 29. Paperwork procedures

instrument, particularly if it helps obtain commitment from those involved, or by those they represent.

Informative meeting

This is a group in which one of the members is the decision-maker and wishes to be informed of various matters before making a decision. Perhaps this is to try out proposals upon an expert audience in order to test reactions. The essence of such a group is that it is really one man who must first seek information from advisors or colleagues before making the decision. This kind of meeting is frequent in projects and can be a most useful instrument provided that certain pitfalls are avoided. There can be a tendency for such a meeting to degenerate into a committee. One reason for falling into this trap is that unless the purpose of the meeting is made known and its objectives clearly stated, some of those present will imagine that it is a committee when in fact it is no such thing. Then, this situation could generate quite serious discords.

Communication meeting

This meeting is the converse of the last. It is a method whereby orders, information, instructions, etc., can be conveyed from one superior to a number of subordinates. Often it is the intention that they will hold satellite communication meetings lower down the line. Analogous to the army procedure 'O' (order) groups. This is a most useful and swift method of conveying information. Its great virtue is that the superior gives out the orders or information and obtains immediate feedback from the members of the group, each of whom can ask questions or otherwise react to what has been said.

The leader (director) can make quite sure that what he or she is saying has been conveyed and, as such, this is the most accurate method of oral communication. Another advantage is that the subordinate people have heard their superior putting over the matter and they can emulate this when they come to do the same themselves. Where people are not skilled speakers this has great advantages, as the armed forces have known for a very long time and have devoted much training at all levels to ensure that these skills are developed.

Discussion meetings

The function of such a group in any organization is to generate, develop, and refine ideas. To work efficiently it is essential that chairs of these meetings should show wisdom and tolerance, and give everybody a fair crack of the whip. There are always those personalities who are hurt by criticism of their ideas, and so must be carefully cultivated by the chairman. On the other hand (and fortunately?) there are people whose flow of ideas is completely unimpeded by criticism; sometimes they seem to be stimulated by it. It is worthy of note here that different professions and walks of life have different methods of training, and that some of these are based almost entirely upon criticism. People in such professions (e.g. architects) are used to criticism and this tends to make them the more forceful and creative in the way in which they put their ideas and how they deal with the ideas of other people. On the other hand, lawyers frequently think that architects are an uncivilized bunch, and so indeed they may be by their own or other standards. Within a 'project group', engineers, contractors, quantity surveyors all have their own cultures with different emphasis on the approach to, for example, science, art, time and money – the individual perception of the importance of which is more highly developed in some cultures than the others. 'Clients' with their different objectives in the project and their many different backgrounds and commercial cultures also bring yet another wavelength to the crowded frequencies through which effective project communication must be achieved.

The efficiency of such a project group depends very largely upon the qualities of the chairman in bridging the gaps caused by such cross-cultural backgrounds and building a successful outcome into cohesive and acceptable messages so as to induce the appropriate action.

Contentious meeting

It ought to be noted that there are some meetings and groups of people who enter meetings with the intention of making it a sort of formal battle. Society has long ago come to terms with the fact that conflict is of the essence of human life, which we share with almost all other vertebrates. Almost everywhere on earth such human fights have been turned into a formal contest in courts of justice. Indeed, the history of the English courts provided for 'justice' through trial by battle, in which each disputant literally appointed his or her champion, and the winner won the appointer's case.

The meaning of this procedure is plain – society believes that if the fight is played according to the rules and all sides are given a fair opportunity to put their point of view as strongly as they can, the result will be beneficial (or at least as beneficial as may be expected in an imperfect world). Although the lawsuit is a fine example we should note that such meetings of groups of people, where some are trying to sell something to the other, is also a contentious meeting of this kind. Whenever buyers and sellers meet and haggle according to the friendly rules the result is thought to be beneficial, because overall it achieves 'the right price'. In the building industry we are accustomed to a great many such contentious meetings, and it is well to recognize them for what they are – very important points of contact between conflicting interests, where the conflicts are resolved in a productive manner – the win/win situation. It is the grit in the oyster that creates the pearl!

It is often said that what distinguishes a group from a mere collection of people is a common objective or intention. But this statement needs a little amplification. The social objective may well be a common one, for example everyone present at a trial in the law courts is taking part in a procedure which will do justice in the end. But, as far as the participants are concerned, each will have their own personal objectives. Some of the people want the prisoner to hang, others want the prisoner to go free, others again have the duty of deciding who is in the right, who is a liar, and so on. The real social objective is in nobody's mind when the group is operating. People are quite properly opposed to each other, but it is emphasized that all of this highly complex procedure must be subject to very carefully enforced rules. The greater the conflict, the greater strictness with which the rules must be applied.

In all of the different kinds of group that we have mentioned above, elements of conflict are bound to arise. Therefore it is for the leader of the group to decide what degree of severity will be imposed by the formal rules of debate, whatever they may be.

Thus the real structure of any group of people is of vital importance. It is resolved by answering these questions:

- what is the purpose of the group?
- what sort of meeting should be employed? (or is a meeting necessary at all?)
- who should attend such meetings:
  (i) regularly?
  (ii) on call?
- who shall chair the meeting?

The answer to each of these questions depends very much on the circumstances. There is no hard and fast rule, and the first task of those responsible for establishing the agenda for and policy to be pursued at the initial partnering workshop will be to decide just how it will be structured and led and how it will set out to communicate to achieve commitment to the projects aims and objectives – this will be most important at the project workshop and for the Project charter. Then it will be sensible to determine what subsequent meetings will be held to sustain the common motivation of the partnering and project objectives. Will there need to be separate meetings for the ongoing assessment, or should it be combined with the project progress meeting? What kind of group meeting is the progress meeting – discussion – contentious – informative – communication?

Let us take an example to illustrate the way in which they can be answered. The director of a company wants to brief himself upon some complex matter before attending upon one of the company's customers. He calls the representatives of various departments to advise him and to discuss the proposals which he intends to make to the customer. In such a case he will obviously wish to sit in the chair, and indeed it is his right to do so.

Let us suppose, however, that in similar circumstances the representative of the company who is going to attend upon the customer is in a much more junior capacity. In such a case ought he to sit in the chair because he is the man who is to be informed and equipped to carry out his duties?

When we have a committee, the chair is much more of a servant than a master, and there is a great deal to be said for having someone not directly involved in much of the minute business of the meeting. Lively spirits find it difficult to be uninvolved, and it is usually wrong to put a person who has strong points to make in the position of being a chair. A good chair must not make the points, and thus the committee loses. If the chair makes the points he or she may not be a good chair.

This chapter has discussed 'groups' in similar terms to meetings. But with a group, its members may be dispersed and not always attending all meetings or face to face discussions with each other. In such cases, great care must be taken with the means and content of communications in order to generate a feeling in the participants that they are part of the group. This, in itself, is a very interesting and complicated subject, but one which, whether within a company or project structure, is vital to successful team work in construction.

Members of the group

When forming groups, for whichever purpose, but particularly in the partnering context we should ask: what can each individual contribute to the group of which he or she forms a part? Generally, all present will usually:

(*a*) speak for a number of others who stand invisibly behind them (the department or the firm)
(*b*) attend in some expert capacity, and have special knowledge which the other people do not possess.

Indeed, the very existence of a group at all is a recognition of the fact that a single individual cannot do everything. In addition to all this, each can usually contribute general observations and criticisms arising out of their own experience and own good sense, and these can sometimes be extremely valuable. How then can we best approach the other members of any group in which we find ourselves?

The following points are suggested to begin with.

- Know the others' work. Embrace every opportunity that arises to learn something of the methods of working employed by other people in any group. Fortunately this is easy, because most people like talking about their work and the way they do it. Any experience that can be gained of another's job – practical experience – is welcomed because it is of such assistance in communicating with people later on.
- Know the people themselves. Everybody has their own personality, and everybody is worth getting to know.
- Know the position which the other holds in their own departments or firms and know who stands behind them (and how far!) – the unseen contributors to any group discussion.
- Know the degree of empowerment given for the project.

Where the group consists of representatives of the same firm, e.g. members of different departments, all of this is relatively simple to do, because it can be carried on conveniently outside the time occupied by group discussions. It is not so easy when all those working in a group, or present at a meeting, belong to different firms or practices with different objectives and different responsibilities and cultures.

How then can this kind of knowledge be achieved? As a start, it will usually clarify the position if you find out what is the legal responsibility of each person, and to whom they owe that responsibility. What have they promised to do? To whom has it been promised? Now you know what he or she *must* do, you can form an opinion of what he or she *may* do. In this way, you can achieve a knowledge of the limits within which the person is able to be cooperative with you. Clearly the attitudes to 'partnering' held by the leaders of all the participating parties – the stakeholders of the project – are vital background information.

Let us take an example. Frequently contractors bemoan the unreasonableness of some municipal treasurer. The treasurer refuses to allow some payment or other which appears to be perfectly reasonable, but which has no legal foundation if the law is to be strictly applied, and is unfavourably contrasted with the private industrialist who can give a direct answer and has only to be convinced of the general fairness of the claim. It is not always realized that the treasurer is a servant of the corporation which is, in turn, very strictly controlled by statute (and the district auditor), and that (in theory) both may find themselves paying out of their own pockets for any monies not distributed in accordance with the strict letter of the law. The limits to this cooperation are very closely set.

On the other hand, private industrialists may be dealing with their own cash on making a commercial decision and do not have to look over their shoulder. If generally satisfied they may spend their money how they wish. The limits of cooperation are then very wide indeed. But, in 1995 with the philosophy and practice of quality management receiving great attention from government and public authorities, partnering should not be ruled out. Why should the district audit office not be brought in as a stakeholder?

In all these cases, the great thing is to get to know the life and work of the other people as swiftly as you can, and time spent upon this kind of social contact at the outset of any undertaking is time very well worth spending. The initial partnering workshop has this role for the project. Furthermore, most people find this process exceedingly pleasant, and so conducive to team building on a project basis.

Finally, since all of this is a two-way process you should go out of your way to facilitate other people getting to know you and the people behind you, and to make your own position clear at the outset.

Behaviour in the group

Each kind of group then has its own special rules. At one extreme stands the committee, whose procedure is well known (or is it?). Brush up on proper committee procedure if you are involved in one, and stick to it. Support the chairman (but do not take over from him) in steering the discussion in the proper way according to the rules. You will find that the committee is then a good deal more efficient.

Other kinds of meetings, though well known, do not have universally accepted procedures. Procedure generates itself, but make sure that the procedure adopted facilitates the purpose of the group as originally organized.

Whether as chair or just a participant, do everything you can to stop one kind of group turning into another, for example an informative meeting turning into a committee.

When you speak in an expert capacity do not use this as a vehicle for making general comment. It is a favourite trick of teachers and clerics to mingle their expertise with comment, exhortation, and matters upon which they are no more qualified to pass judgement than anybody else. This kind of approach is inefficient and leaves a bad taste amongst those being exhorted!

Where experts are being asked to give expert advice they should do so. Their general observations may be requested and these should be kept separate. For example, when the manager of a pre-cast concrete factory gives his views on the work of his own department these are naturally listened to with considerable respect. The respect stems from his expert knowledge of the details of concrete production, which is little known to the others present. If later on he offers some 'wise' comment on a matter coming from the design department the respect with which this is received will stem not from this expert knowledge but merely from his age and general experience of life. The weight attached to the two different statements can vary enormously, particularly if others present have equal or greater knowledge of the design factors.

In contentious meetings it is not right to expect people to trap themselves into giving away their rights, or those of the firms they represent in off-the-cuff discussion. This is too frequently done, and the effect of it is to slow down communication and bring it virtually to a stop. It will also help to destroy the pursuit of a common objective and a win/win situation.

The lawyer's 'without prejudice' discussion is a useful and civilized procedure to get matters moving when people might otherwise tend silently to 'stand upon their rights'. Perhaps this should be used more often, at least until there is genuine commitment throughout the team on the common objective – a quality project finished on time and to agreed cost, where everybody wins – and then their overall costs at least are therefore reduced.

Finally, agenda are of great importance and can contribute to success, even in the most free-wheeling types of group. If it is remembered that representatives coming to a meeting are speaking for numerous people behind them, they cannot be expected to deal with matters of which they have no notice. In the very complex arrangements arising in building projects, too much is often expected of those who attend a site meeting. Get the queries on to the agenda and publish the agenda in good time. In itself this discipline will help to remove many irrelevancies and it will also help to ensure that matters are properly cooked before they come to discussion – and a great deal of time will be saved by the process.

Many of these points, and many other procedures not touched upon, are really just common-sense and an awareness of human strengths and weaknesses, but the procedures cannot be applied successfully without knowledge of the other people in the group and their own sub-motivations. Without this knowledge trust is unlikely. It is this that must be developed by commitment, communication and good leadership to create a commonly motivated team that will together achieve the common objective.

Bringing the specialist and smaller sub-contractor into the partnering process

> When you are skinning your customers you should leave some skin on to grow so that you can skin them again.
>
> *Nikita Krushchev to* British Business News, *May 1961*

In Chapter 4 we looked at the possibility of conditioning the sub-contractor to the idea of partnering at the earliest possible stage of the project when pre-qualification through a project-specific audit was undertaken. This has the potential for double benefit in that if convinced, the sub-contractor should be more keen to participate and if this is pre-bid – his bid should be lower – reflecting that partnering and TQM should produce lower costs in overheads, materials and manpower – the reduction by 30% as postulated and sought in the UK – and experienced in successful projects in the USA and Australia.

In any event the specialist should have been educated (and learnt) that TQM is not about paperwork systems, standards and third-party certification. It is about improving the competitiveness and performance of the company as a whole and of the ability to provide profitably the product or service – or both – that its customers require.

For the trade contractor this means improving his or her trade skills as well as delivering the service on time and to the agreed cost limits.

For the specialist sub-contractor it will include buying and manufacturing, delivering and maybe assembling and working around other specialists, with all the planning, organizing and controlling functions of management that this entails, whilst at the same time establishing and maintaining good working relationships with others over whom he has no control. This will require good 'upwards' and 'sideways' management.

For small firms where the principal is in close touch with all his or her workforce and with everything that goes on day-to-day, a few paperwork procedures and some small definition of duties may be all that is necessary, but they should be done and in relation to the overall project structure. The philosophical changes necessary to develop TQM on a project basis may also seem to be a matter for others than the principal sub-contractor. But, the cynicism borne of long exposure to the wiles and whims of the architect, engineer and general contractor will require a lot of attention before the sub-contractor can even begin to believe in the cultural changes that will be needed to improve his or her position at the end of the line – as everyone's scapegoat – or that this will ever be brought about.

But, TQM is a holistic approach to providing customer satisfaction, and is far more relevant to the small (sub-) contractor than just the rigid discipline of a quality standard. In small traditionally-based sub-contractors it is less likely than with the design professions and general contractors that there has been *any* formal management training. It is highly unlikely that the sub-contractor's management will have been exposed to the philosophic approach of Deming, Juran, Taguchi, TQM, Maslow, or McGregor, although they may feel that there is the need for culture change on the construction project and in all the parties of the team involved before them if the project is to achieve its success!

Internal customers

The biggest change that TQM offers related to sub-contractors in the construction process is to put them in the role of 'internal customer' of the general contractor.

The 'internal customer' concept is that within every process everyone has a customer.

In manufacturing industry the internal customer is seen as the next group down the line. For example, on a motor-car assembly line the group who place the chassis has for its customer the group that places the engine on it, that group's customer is the group that fit the bonnet and axles; further customers fit the wheels and then the seats and then fix the trim, until at the end of the line the cleaning group is left with little to do before the ultimate external customer proudly drives the finished vehicle away – perfect in all respects. Even more simply, when a boss dictates a letter to the secretary, the secretary is his or her customer and should be treated as such. Conversely, when the secretary delivers the letter with its enclosures and envelope neatly presented and in the right order, he or she does so to a customer rather than to a boss.

The implications of this concept in relation to a construction project are vast and dramatic. The last group in the construction of a building are the painters and decorators or floor-layers, generally specialist sub-contractors. The total quality management philosophy makes them all internal customers of the general contractor!

The general contractor in his turn is the customer of the quantity surveyor for the bills of quantities. The quantity surveyor is the customer of the architect and the various design engineers related to the drawings and information on which the bill of quantities is based. If this is extended further it places the design team as the customer of the project manager, if there is one, or the client!

The philosophy behind this concept may be no more than the age-old 'do unto others as you would be done by', but is the culture of the construction industry so different that the message is invalid? Team-working with the client as part of the team in a genuine partnership to achieve project objectives can and does work, given the right leadership and contractual relations, all part of project quality management.

The better clients, the better professionals and the better contractors know this already, and enjoy repeat and continuing business to their mutual satisfaction. But with the unique project orientation of most construction projects, repeat business is far from the norm, and cultures have developed across the industry that require substantial cultural change from client and contractor, design professionals and specialists alike.

Most sub-contractors will need a great deal of convincing that partnering really will develop the internal customer principle as far as they are concerned. It is therefore the client who must demonstrate by performance, from the first contact with the sub-contractor through to the initial workshop, that it is *going to work on this project.* It is the client who can do most to provide the sub-contractor with the necessary conviction – perhaps with financial evidence of success on other projects as case studies.

7 The role of the facilitator and contract management adjudication in problem resolution

Since wars begin in the minds of men, it is in the minds of men that the defence of peace must be constructed.

Constitution of the United Nations Educational, Scientific & Cultural Organization, 1964

Substantial parts of this chapter have appeared in Managing Construction Conflict, *written in 1987 as a result of Polycon's experience in developing alternative dispute resolution techniques from formal arbitration under the 1950 and 1979 Arbitration Acts, and also in Chapter 13 of* Total Quality in Construction Projects. *Both were quoted in the Latham report.*

Since the concept post-Latham is proposed to be the subject of legislation, for which, at the time of writing, parliamentary time is being set aside for a construction contracts bill, it is included here – in essence as previously written but with particular application within a partnering concept and charter.

That it applies, and can make a contribution to the overall improved performance of construction projects without either legislation or a partnering charter, is self-evident, having been used, if only occasionally, by Polycon as a dynamic tool to facilitate dispute resolution for some 15 years. Nevertheless the project climate created by partnering creates synergetic self-support between the two ideas – providing the resources are developed by training adjudicators as facilitators, and perhaps also of professionals in quality and its facilitation in the principles and practices of more formal arbitration under the acts.

Latham report – Recommendation 26

Adjudication should be the normal method of dispute resolution.
and at paragraph 9.14 :

(*a*) A system of adjudication should be introduced within all the standard forms of contract (except where comparable arrangements already exist for mediation or conciliation) and that this should be underpinned by legislation. I also recommend that:

- there should be no restrictions on the issues capable of being referred to the adjudicator, conciliator or mediator, either in the main contract or subcontract documentation;
- the award of the adjudicator should be implemented immediately. The use of stakeholders should only be permitted if both parties agree or if the adjudicator so directs;
- any appeals to arbitration or the courts should be after practical completion, and should not be permitted to delay the implementation of the award, unless an immediate and exceptional issue arises for the courts or as in the circumstances described in (4) overleaf;

and at paragraph 9.13 :

(*b*) Very senior judges, in the High Court in the UK, have stressed that holding up the flow of cash is bad for the construction industry. Lord Justice Lawton, in Ellis Mechanical Services vs Wates Construction Limited, 1976, 2BLR 57: 'The Courts are aware of what happens in these building disputes; cases go either to arbitration or before an Official Referee; they drag on and on; the cash flow is held up . . . that sort of result is to be avoided if possible'.

Extracts from *Constructing the Team* by Sir Michael Latham.

It is implicit in total quality management that the traditional attitudes that have prevailed in the management of construction companies and projects must change. In construction the teamwork for TQM will not survive the difficulties experienced in construction on site when these problems arise without a means of continuing the harmony between parties in a conflict situation. Contract management adjudication is a new concept developed to continue pre-site harmony in the heat of construction endeavours. Management techniques must provide the procedures and practices as tools to give effect to the principles, which evolve, essentially, from a philosophy. These tools are set out as the means of retaining harmony without prejudice to any of the parties' ultimate legal rights of redress. They will add value and reduce the cost of quality by eliminating or subsequently reducing the costs of dispute and subtracting the negative value of legal fees.

We have seen in earlier chapters that the project quality plan must start at the top – and at the beginning – with the client, or his or her representative and project manager, if the plan is to be fully effective. Let us assume that this has been done, and the necessary goodwill has been established initially through a workshop to create an effective team spirit towards achievement of the common objective – the common goal of customer satisfaction. A quality building will still only be achieved in practice, on site, where the unexpected can be expected. So, frequently changes must be made not only to the project plan but perhaps also to the design, or even the requirements for the design. It is from this that claims can arise far beyond those provided for in provisional or pre-contract sums, notwithstanding any savings that may be developed through value engineering.

This is the stage in the overall process when claims rapidly lead to disputes, and to protection of self-interest by the many parties involved. The team spirit can quickly be shattered on unexpected subterranean rock! But, the partnering spirit must prevail if the project is to be successful. This is the place for a management technique evolved from a combination of the philosophies of arbitration with those of TQM into contract management adjudication (CMA).

TQM

TQM should begin, not just with the construction phase, but at the time the brief is developed and passed to the architect/designer. Then, the procedures will eliminate many of the communication problems that beset the project, delay its completion, increase its costs and lead so often to the depressing pattern of claim, counterclaim, dispute and ultimately litigation, or arbitration, which so beset the industry and benefit no-one except the lawyers!

It is really the extremely high costs of *poor* quality in the overall construction process that will be eliminated by ensuring that quality management systems are used on the project.

Such quality management systems will improve performance, reduce costs and increase profitability. But because all construction projects have a built-in potential for dispute through conflicts of interest, there is need for a further step to keep the industry 'out of the courts', which alone will have a substantial effect on management stress, company morale, public relations and profitability.

Multi-disciplinary projects have grown larger, more complex and more prone to confusion and conflict between the parties. Legal dispute has been the growth industry of the last decade.

Good management, developed through the TQM process, is undoubtedly the preventive medicine of dispute; but differences do arise, and good management of the dispute procedure then becomes an even more vital requirement.

Alternative procedures, starting with TQM, to resolve technical problems are long overdue. This is where techniques involving *adjudication* can be of value, using the benefits of arbitration, recognizing that the technical content of a dispute is at least as important as the contract clauses under which the project has developed, and giving the adjudicator greater freedom to intervene than is customary in English arbitration.

The adjudicator's role is akin to that of a cricket umpire. Normally the umpire intervenes only when one of the sides appeals for a decision as to whether a batsman is out, but the umpire also has other duties; to observe 'no-balls' and offences against the rules that might otherwise go unobserved or cause dispute. Adjudication techniques are equally applicable to contractual disputes concerning materials supplied, or work done, and are particularly suited to claims where an insurance company is also involved indemnifying and so backing one (or sometimes both or several) parties.

For the specialist sub-contractors not only offset of their claims but all aspects of the conduct of the project, should be covered by adjudication – and can be, if these provisions have been built in before the documents go out to tender to both main and sub-contractors – as part of the project quality plan.

It may be better for the advancement of the partnering process on specific projects if the concept is developed at the partnering workshop and if the adjudicator, a competent facilitator, is also present. The process is then owned by the stakeholders rather than imposed upon them by the statutory provisions in the contract form.

Thus the tendency to regard litigation, arbitration or adjudication as necessary evils to be invoked in order to settle a dispute which is already under way is eliminated in favour of a dynamic value adding technique. There is, however, a third role for adjudication and that is as an extension to the TQM of a contract. Adjudication is a powerful management tool which should be used early, positively and dynamically to improve performance in an area which currently wastes that most important resource—management time.

Adjudication is a cheaper and quicker means of settling disputes or claims; and, better still, as the last stage of TQM, as a means of reducing the chances of dispute ever occurring. This method of resolution retains the advantages of arbitration for technical disputes – where the arbitrator is chosen for his or her knowledge of the matters in dispute and can let personal expertise influence his or her judgement.

If the adjudicator/facilitator has been present at the initial partnering workshop he or she will have the added and substantial advantage of being immediately *au fait* with the general subject area, the probable problem areas, and the personalities, and so be able to speed the award process and reduce its costs. These are substantial value adding benefits over both arbitration and litigation.

The problems of litigation or arbitration on technical issues are that:

- the action is normally brought long after the event, when witnesses' memories have dimmed

- the case often turns on expert opinion as to the state of the art several years previously
- the vast quantity of documentation takes a great deal of time for everyone concerned with the action to read and absorb, adding greatly to the cost
- the English adversarial system of litigation necessitates solicitors, advocates and experts being retained by both parties, or, in the not infrequent case of seven or eight defendants, by all parties to the action; all adding to the time and cost of reaching a conclusion
- sometimes, too, the outcome is one on which the judge and the lawyers agree, but the parties and their technical experts have great difficulty in reconciling with their knowledge of the technical facts.

Arbitration can, and frequently does, provide a better solution than litigation for technical disputes. Adjudication advances the time of solution to the time of the incident.

Recent years have seen several developments of adjudication technique. The British Property Federation has formulated a project management system in which the role of an adjudicator is defined. In the USA a system of adjudication has been developed to resolve disputes by 'mini-trials' and 'pre-trial reviews' where the main issues are examined by lawyers and technical experts, but by informal hearing of predetermined and limited duration. All these techniques have the objective of a quicker, cheaper and more business-like resolution of the dispute.

So greater benefit can be obtained on long-term projects if adjudication is established by the parties at the outset, and this should be built into the TQM quality plan. Observations can then be made on the contract, its documentation, and on preventive measures throughout its operation by the adjudicator. This technique is known as contract management adjudication (CMA).

The purpose of contract management adjudication

The purpose of CMA is to quickly resolve disputes, to improve performance and reduce costs to all parties throughout large or multidisciplinary projects. It recognizes that disputes are sometimes inevitable and that their speedy and economic resolution will be made by objective and impartial decision by experienced professionals who have not been involved in the design, development or formulation of the contract. It also recognizes that the authority of professionals and contract management within the project must not be disturbed.

The adjudicators, at the start of the contract, examine the project to identify gaps and overlaps in the contract documents, contract organization and (to the extent that it exists already) the project quality management system. This system should recognize that:

- disputes in such activities are sometimes inevitable
- speedy and economic resolution can be made by objective and impartial decision by experienced professionals who were not involved in the design, development or formulation of the contract
- the authority of the professionals and the contract managers must not be disturbed.

Traditionally, forms of contract have named the architect or engineer as quasi-arbitrator. Yet, they already have a contractual relationship with one of the parties. Further, likely subject-areas for dispute result from their design or organizational decisions. A designer may have omitted data from his or her specification on which contractors based contractual promises, or they may have made what turn out to be wrong decisions, costly for the contractor to correct.

CMA

CMA – a project management technique, whether within a partnering charter or not – will:

- reduce avoidable delays during the progress of the contract
- prevent dispute immediately a situation arises that might damage one of the parties or prevent them from proceeding with the work
- reduce the costs to all parties involved in the dispute
- eliminate the 'judge in own cause' and conflict of interest for the professionals whose earlier decisions become an issue
- record factually and objectively all issues dealt with, and give expert and impartial opinion should any party still wish to contest them later and in court
- monitor and audit the project management system.

CMA combines concepts of post-contractual arbitration with organizational project management (see Fig. 30).

Initial appraisal

Problems occur most frequently on complex projects if lines of communication and the lines of contract are not coordinated at the onset (see Fig. 31). Therefore an initial appraisal is carried out by examining the project to identify gaps and overlaps in:

- the contract documents
- the contract organization
- the project quality management plan.

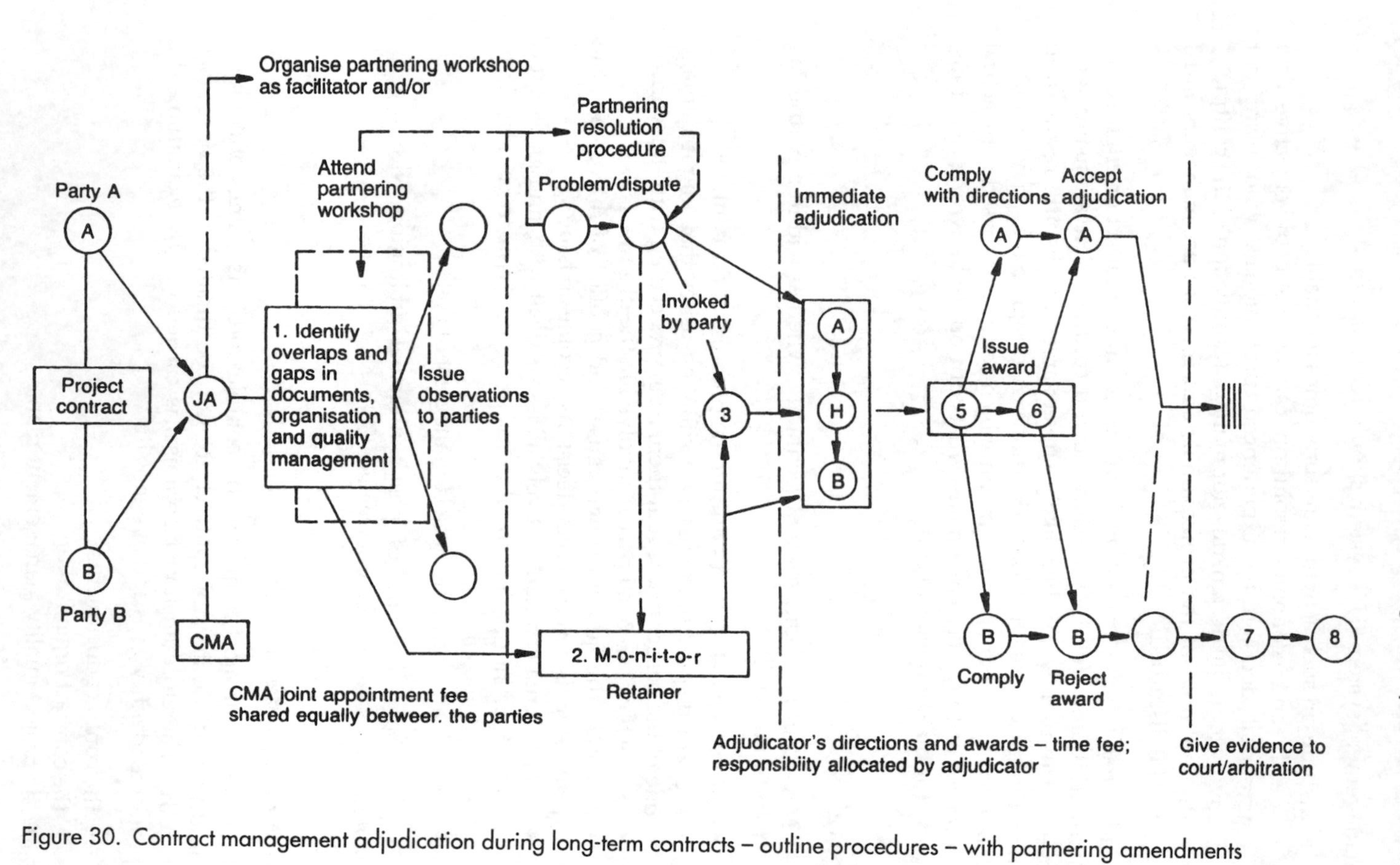

Figure 30. Contract management adjudication during long-term contracts – outline procedures – with partnering amendments

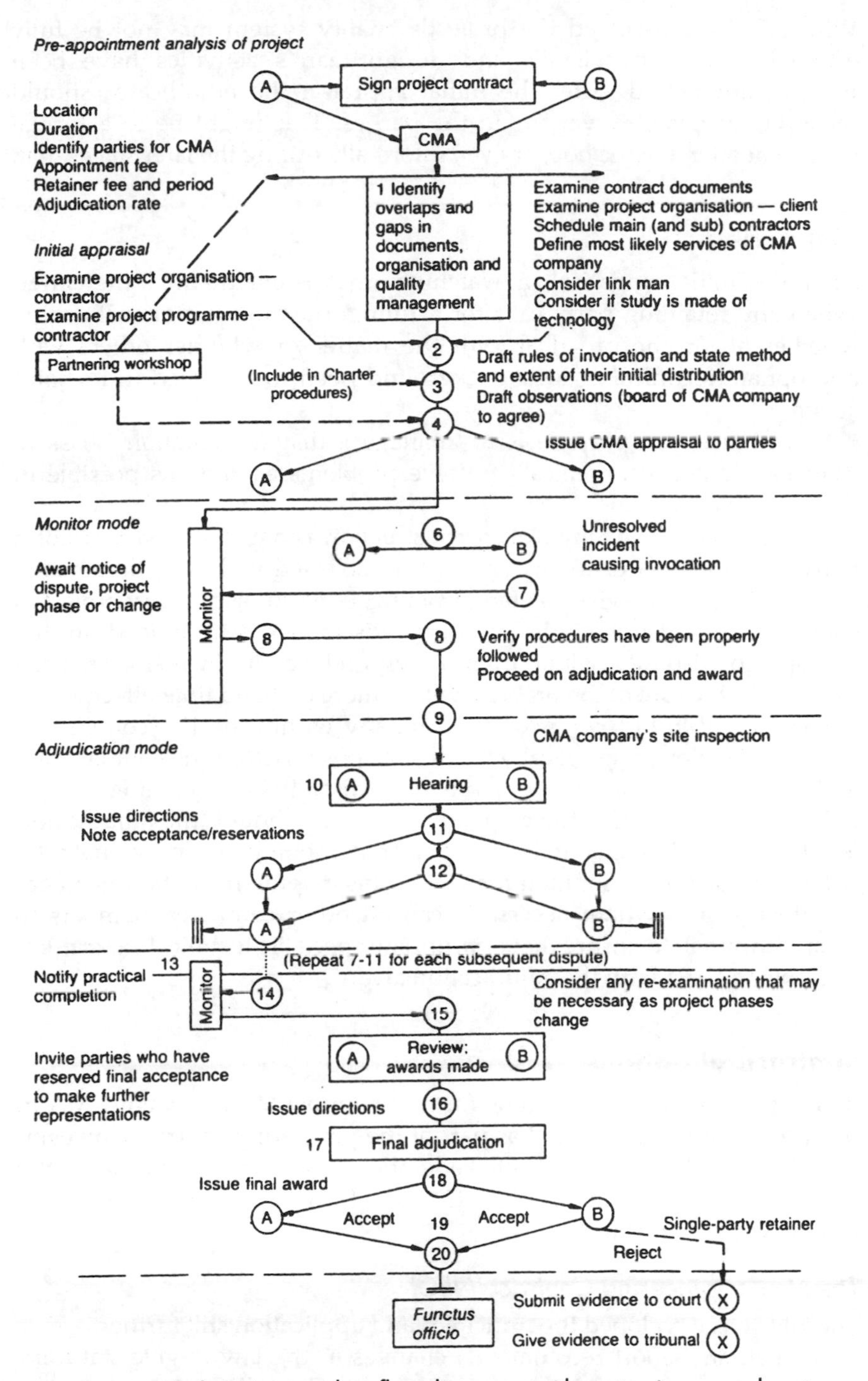

Figure 31. Detailed CMA procedure flow diagram – with partnering amendments

Where TQM is involved the project's quality system may not be fully effective unless the client's and design team's activities have been properly integrated. After the initial appraisal the adjudicator should make his or her observations to the parties. This might be in a formal document after a workshop, or even informally during the later stages of it.

Monitoring

After the initial appraisal, a watching brief is maintained thereafter. When any retaining party calls for a ruling, the adjudicator will decide whether he or she can deal with the matter or whether others with appropriate technical or legal background need to be called in for joint action.

In a phrase, the purpose is to 'manage a dispute situation' so as to dynamically and economically resolve problems as early as possible in the chain of causation.

Whilst in principle, any skilled construction professional who is not a party to the project or to any contracts established within its framework, could fulfil this role – given suitable training. In multi-disciplinary projects a company possessing all the technical disciplines and trained in the principles of dispute resolution, able to respond technically no matter what the technical nature of the problem will be more cost and time-effective.

Polycon AIMS Ltd, an associate company within the Polycon Group, has been developed specifically for this purpose. Within the company are adjudicators from all the construction disciplines, including legally qualified members. It is the company that is appointed 'the adjudicator' (see Fig. 32), and as a separate legal person it appoints the individual who will be the dispute resolution manager. This dispute resolution manager can then, if and where necessary, call on other company members to form a 'tribunal' to ensure that appropriate specialist technical expertise is available to assist with the adjudication award.

Contractual options

It is not essential to have a new form of contract to implement CMA in practice, it could be included as part of the partnering charter. However, already The British Property Federation contract system and the Association of Consulting Architects form of contract incorporate clauses dealing with adjudication.

The New Engineering Contract (NEC)

The NEC has developed the principles and application still further.

The Latham report recommends changes in the law to give statutory backing to a proposed new standard form. The 'NBC' (New Building

Polycon A I M S Ltd. Adjudication and Arbitration

POLYCON MANAGEMENT

70 GREENWICH HIGH ROAD • LONDON • SE10 8LF TEL: 081 691 7425 FAX: 081 692 9453

APPOINTMENT AS CONTRACT MANAGEMENT FACILITATORS/ADJUDICATORS

WHEREAS:

1. On the .... day of.. 19 ....

   a contract was entered into for the purpose of:..

   ....

   between ....

   of ....

   and...

   of ....

and

2. The parties require the services of Polycon AIMS Ltd. to adjudicate during the course of this contract in the event of a dispute arising.

and

3. That in the event of such dispute arising both parties to the contract undertake to:-

   i) abide by and comply with the rulings of the Adjudicator throughout the period of the contract;

   ii) to pay the costs of the adjudication service or to ensure that it is paid by the party whom the Adjudicator assesses should meet the fees;

   iii) to give notice within the prescribed period after each adjudication and at the end of the contract if they intend not to accept the Adjudication as final and binding upon them in all respects;

   iv) in the event of any matter related to or arising out of the main contract which shall become the subject of litigation or arbitration after the completion of the contract) agree to the submission of the adjudication awards as evidence in action.

4. That this appointment shall be irrevocable by either of us without the written consent of the other prior to the completion of the project which is the subject of the contract.

5. That the appointment of Polycon AIMS Ltd. as Facilitator/Adjudicator will commence on signing and will continue until completion of the contract and/or the payment of the Final Account, whichever be the sooner.

Signed by the Parties:-

Signed: .... Party A

Signed: ____ Party B

Date: ______________________________

Company Registered in London No. 1193380 - Part of The Polycon Group

Figure 32. Appointment of adjudicator (or adjudicator/facilitator)

Contract) which, like the NEC, would provide for adjudication, would be binding during the progress of the works but subject to review if any party so requires at the end of the contract.

Whatever arrangements are made by the partners for dispute resolution in the partnering agreement, which should quickly move any difference or dispute to the highest level in the participating organizations, there should always be the provision for CMA, preferably by an on-the-spot 'referee', and preferably one versed in the particular areas of the problem – hence the benefit of nominating a 'legal person', i.e. a company that can provide all the requisite technical skills!

As referred to above, Polycon AIMS Ltd have developed a service which gives practical and highly cost beneficial effect to these principles. It was developed separately but overcomes the disadvantages that exist with these and other adjudication techniques.

Polycon AIMS' adjudication and arbitration procedures stem, as does all arbitration, from the authority of the parties. They embrace concepts of the powers given to arbitrators under English law and some of the more normal practices which flow from continental legal practices.

CMA in particular has been developed to be applicable to, and usable with, any and all of the standard forms of construction contract. It does not require any change to be made to the standard forms and thus can be implemented immediately, but it is obviously better to build it fully into the project plan from the start.

Polycon AIMS can be appointed as CMAs any time after the main contract has been signed, whether or not a difference has already arisen. But it is better that provision should be made in the specification prior to the signing of the main contract, in which case the following clause should be included:

> In addition to/substitution for the provision for arbitration under Article 4 of the Recitals, Contract Management Adjudication procedures shall apply to disputes or differences arising out of or in the course of the works.
>
> It is also intended that Polycon AIMS Ltd of 70 Greenwich High Road, London SE10 8LF, shall be appointed jointly by the employer and contractor as adjudicator to resolve any dispute or difference that may arise during the progress of the works between the employer or his or her architect, engineers, quantity surveyor or supervising officer and the contractor or any sub-contractors, to issue directions for the continuance of the works under these conditions but without prejudice to the right of any party to seek formal arbitration if they do not accept the adjudicator's award as final and binding upon them all at the end of the contract.
>
> The contractor will be required to incorporate the provision for

adjudication in any sub-contract for the works and for the supply of items for the works whether these are from nominated persons or otherwise.

In a partnering contract, where this has been the client's intention pre-tender, the wording will differ. To the first paragraph should be added:

> At the partnering workshop procedures will be developed that will provide the initial mechanisms for problem-solving on site, and off it before the adjudicator's award is invoked, etc.

See Table 7 for the issue resolution system used on the Macquarie University project.

Probably the external adjudication process should come as the penultimate step as drawn in Fig. 30 – and would replace the top management final step.

Following the Latham report the use of an adjudicator as part of the contract management process is intended. This will apply whether the contract has a partnering clause or not. The project will be less susceptible to even adjudication during the progress of the contract if it is within a partnering arrangement and for which a charter has been prepared. Adjudicators who are competent to act as facilitators and assist with the organization of the partnering workshop can then be appointed to the dual role as facilitator/adjudicator and the more successful they are in the former the less need will there be for them in the latter. In which case the appointment form as adjudicator could be amended as follows, inserting:

'and 2. It has been agreed between the parties that a partnering process shall be used, an initial workshop will be held and a project charter produced.

and 3. The services of Polycon are required to assist with the organization and conduct of this workshop and to attend in the role of facilitator, and no later than seven days after the completion of the workshop to issue their observations on the contract documents and on the partnering charter in relation to the problem resolution procedures proposed in the charter'.

Reverting to the original:

'and 4. The parties require the services of Polycon to adjudicate during the course of this contract in the event of a dispute arising.

and 5. That in the event of such dispute arising both parties to the contract undertake to:

Table 7. Issue resolution system on Macquarie University

Sub-contractor	Theiss contractors	Principal/ client	Superintendent/ engineer	Testing authority	Designers	Time frame
Foreman	Foreman		Inspector	Field technician		2 hours
General foreman/ supervisor	General foreman/ supervisor		Senior inspector	Field technician		4 hours
Project engineer	Project/quality engineer		Project engineer	Laboratory manager	Design engineer	1 day
Area manager	Project manager	Principal representative	Project manager	Laboratory manager	Design manager	1 day
Area manager	Construction manager	Principal representative	Area engineer	Senior geotech. engineer		2 days
Area manager	Area manager	Principal representative	Regional director	Senior geotech. engineer	Design manager	2 days
Managing director	General manager	Executive director				3 days
Managing director	Managing director	Executive director				5 days

Principles of process:

1. Resolve problem at lowest level.
2. Unresolved problems to be escalated upward by both parties in a timely manner, prior to causing project delays and/or costs.
3. No jumping levels of authority.
4. Ignoring problem or 'no decision' is not acceptable.

(*a*) abide by and comply with the rulings of the adjudicator throughout the period of the contract
(*b*) pay the costs of the adjudication service or to ensure that it is paid by the party whom the adjudicator assesses should meet the fees
(*c*) give notice within the prescribed period after each adjudication and at the end of the contract if they intend not to accept the adjudication as final and binding upon them in all respects
(*d*) and that in the event of any matter related to or arising out of the main contract which shall become the subject of litigation or arbitration after the completion of the contract to the submission of the adjudication awards as evidence in such action'.

Advantages of technical adjudication by the company

The advantages of technical adjudication by the company (rather than by an *ad hoc* arbitration/adjudication tribunal) are as follows:

- it is the company that makes the award
- the cost of obtaining the award is far less than with litigation or arbitration, and can be predetermined
- all aspects of a complex dispute are dealt with through the multidisciplinary composition of the company's expert and arbitrator members. These reflect the total range of disciplines which are involved in the conception, design, development, cost negotiation of a project – the contract – and its execution. If there is a particular area of (narrow) expertise required to investigate a particular point of conflict which is not contained within the company's expertise a technical assessment is obtained from a neutral specialist – if necessary the best on a worldwide basis
- this eliminates the multiple costs for sets of experts for each or all parties. Whatever expertise is required is contained in the adjudication tribunal and then is built into the award
- the parties can reserve their options and so do not have to commit themselves to either final acceptance or rejection until the end of the contract when, if not objected to, after a pre-determined period it becomes final and binding. But whether within a partnering charter or not the parties will be bound by the charter (and probably the contract) to give effect immediately and to implement in every way the adjudicator's award until the end of the contract
- where an award is made by the Polycon (Ltd) Company, tribunals are scrutinized by other directors and countersigned before issue to ensure that the reference has been fully and properly discharged in accordance with the company's contract of joint appointment and that there are no errors on the face of the award, or in its preparation, or the conduct of the reference.

Conclusions

Few contractors, architects or clients are yet in a position to ensure that a total quality management plan and organization system will be implemented throughout the project's life by a complete team, each part of which is committed to total quality management in their own sphere of activity. However, even if, and when, this is the case there will still be the need for effective means of maintaining the motivation for, and commitment to, the project – rather than the individual interests of various parties involved in the nexus of contracts that has been created when post-contractual conflicts of interest arise.

CMA can play an important part in reducing the effects of these conflicts and should be seen as the final link in the total quality management chain, and one which will give 'partnering' its final binding cement to 'win/win'. This ensures that at the end of the project no-one will feel that going on to formal arbitration (or litigation) is necessary, sensible and certainly not likely to be cost-effective for any of the stakeholders.

8 Case studies of successful partnering – USA, Australia, UK

The race is not to the swift, nor the battle to the strong.

Ecclesiastes, Ch. 9, V. 11

The evolution and practice of the partnering ethos are considered and a series of case studies covering award-winning projects in the USA are examined.

Some partial successes and some successes in Australia are also examined, together with their reasons, alongside some early work in the UK, with reference to the common economic and legal background in these two countries.

The clear conclusions are that the greater the commitment at top level, the more attention given to education and communication with a workshop developing a charter, the greater will be the success in all measurable performance and activities in quality and in time, cost and in human relationships and satisfaction to all participants.

Developing the framework of partnering is generally accredited to Charles Cowan, a Colonel in the US Army Corps of Engineers, and later the Director of the Arizona Department of Transport (ADOT). He was certainly responsible for the exciting growth of its use in Australia following a lecture tour he made in 1992 at the invitation of the Master Builders Association of Australia.

Du Pont Engineering had made earlier efforts at partnering in the private sector with Fluor Daniel as their contractor partner and although both are now firm advocates of partnering, Fluor Daniel still prefer to use their cost reimbursement controls. Cowan began to experiment in 1989 and most of the case studies have the work of Cowan and his colleagues in the army corps at their root, perhaps because cost-plus contracts are not permitted in US government agencies.

In the Portland District of Arizona projects that have been partnered range in value from $5 million to $140 million and have achieved:

- 80–100% reduction in cost growth
- virtual elimination of time growth on project schedules
- two-thirds reduction in paperwork
- millions of dollars saving through value engineering
- no outstanding claims or litigation
- significant improvement in safety records, and have put 'fun' back into the process.

The benefits, as will be seen from the studies which are based on those listed as winners of the Marvin M. Black Awards, are not limited to government owners, nor to any particular kind of project.

As background, the US Corps of Engineers had documented that contract claims in the 1980s had escalated by 200% and averaged over $1 billion annually. This led to Cowan seeking a better way to deal with problems identified by the USA Construction Industry Presidents Forum as 'litigiousness with all its negative impacts as a top industry problem'.

Cowan decided that the tragic deterioration of one of the USA's most important industries was the result of a shift of focus from operating in the spirit of the law to abiding by the letter of the law.

In the current market in the USA great efforts, time and money are invested by all parties to minimize risk. Under the current adversarial framework that dictates much of the construction work done in the USA, the potential for cooperation on a job had been eroded to a point where it sometimes appeared that the goal was no longer the successful completion of the job, but rather to ensure that any failure, cost overrun or schedule conflict can be blamed on another party and successfully defended in court.

Cowan's work was picked up by the Associated General Contractors of America who in 1992 instituted the Marvin M. Black (who was the President of AGC at the time) Award for Excellence in Partnering (see Appendix 3 for the criteria). There are eight awards in each year. In 1993 out of 56 entries they were made to:

1. Eielson Air Force Base, Alaska — $26 million
   (Public) Stakeholders 33
2. USAA South-east Regional Office, Tampa, Florida — $60 million
   (Private) Stakeholders 74
3. Bonneville Navigation Lock Cascade Locks, Oregon — $140 million
   (Public) Stakeholders 9

4. Chevron Wastewater Treatment Project — $100 million
   Port Arthur, Texas
   (Private) Stakeholders 5
5. The Lacey V. Murrow Replacement Bridge — $88 million
   Seattle, Washington
   (Public) Stakeholders 32
6. Menasco Aerosystems Manufacturing Plant — $3.8 million
   Expansion, Fort Worth, Texas
   (Private) Stakeholders 30
7. Mayflower Homes Health Centre, — $3.8 million
   Grinnell, Iowa
   (Private) Stakeholders 39
8. Morris Sheppard Dam Stilling Basin, — $3.5 million
   Possum Kingdom Lake, Texas
   (Public) Stakeholders 4

As can be seen, they have covered a variety of products ranging from domestic buildings at $3.8 million to large civil engineering products at $140 million, whilst the number of stakeholders involved ranged from 4 to 74, and geographically they were spread from the Gulf of Mexico in the South East to Alaska in the North West!

Neither are these isolated examples of success.

The Sundt Corporation – builders since 1890 – had by the autumn of 1994 partnered more than 30 projects. Their experience has, according to President J. Doug Pruitt, varied greatly from project to project but on all but one project they have had positive results, and on those contracts they have had no claims or litigation. The partnering process opened the door and provided the forum to develop win/win attitudes among all stakeholders and improved the way they conduct business at their job sites.

In 1994, a further seven public and one private project out of 49 entries received the Marvin Black Excellence in Partnering awards. These ranged from $7 million to $85 million in size (£4.3 million to £40.6 million) and included a coal fired power plant, a research and development faculty, a water reclamation facility, military housing, a port of entry and highway overlay, and a highway interchange project.

Their total value was $168.1 million (£105 million). All came in on or ahead of schedule, on or under budget, with savings totalling $3.3 million through value engineering and incentive bonuses. No claims were filed on any of the projects. The number of stakeholders involved in these projects ranged from 5 on the $30.6 million water reclamation scheme, to 42 on the $21.6 million coal fired power plant.

The first eight case studies (which are based on the eight award winners of the Marvin M. Black Award for Excellence in Partnering in 1993) are included at the end of the chapter without further individual comment – as none are needed. But by their range they illustrate that it is not the kind of project that makes it suitable for partnering, but the attitude of the people involved in it that makes for success. There have been 51 entries (worth $1 billion plus) received for the current year's award. From an AGCA survey, 675 firms are definitely planning to implement partnering, and a further 600 may do so in 1995/96.

But, is the USA a different culture, presenting a different climate (well, maybe in Alaska) than exists in Britain as we approach the millennium? Can these examples be translated into UK practice?

Before attempting to answer these questions let us look at some Australian examples – one at least of which offers perhaps the best lessons from partial success – or is it partial failure?

Background to Australian case studies

The Australian construction industry mirrors in many ways that of the UK. It has experienced a severe recession in recent years. It has a confrontational approach; conflict, dispute, arbitration and litigation are rife, and the Gyles Royal Commission into the productivity of the industry in New South Wales addressed many issues similar to those which were the subject of the Latham study and report, and with many similar findings.

The Commission learnt of the US experience of partnering and invited Charles Cowan of the US Corps of Army Engineers, then with ADOT, to explain the concepts in a series of seminars in Australia in 1992.

These had electrifying results, and since then:

(*a*) partnering has been included as a recommended practice in the Master Builders of Australia's Code of Practice
(*b*) the Gyles Commission promoted its use in the Science and Technology Building at Macquarie University
(*c*) it is used as the first case study in *Strategy for Excellence* by the CIDA* on the North Dandeloup project. This project was the winner of the 1993 Australian FMBA Successful Partnering award
(*d*) partnering has also been used on other projects, including two reviewed here:

- the Queensland Sunshine Coast Motorway – Stage 2 (10 CE)
- the Green Island Redevelopment, an important ECO project† (10 B)

*The Construction Industry Development Association – the government agency for reform of the construction industry.

†The author is grateful to Roger Quick of Gadens Ridgeway, Lawyers of Brisbane, for permission to draw on his work, and where appropriate it has been indicated where the comments made are his rather than the author's.

Clearly the concept in Australia has its origins in the successful applications in the USA but, as can be seen from the examples which follow, they have been adapted to local conditions – sometimes with greater success where all the fundamental principles to bring about the culture change necessary for total success in quality management have been observed.

Legal constraints

There is much that is similar to English law in the Australian code, and until recently Australian construction lawyers, like many of their English counterparts, have said that knowledge of the terms of the contract and the legal issues that they raise is vital to the proper administration of a contract and that an equally essential part of the proper administration of a contract is an understanding of the legal bases upon which contractors are entitled to construct claims.

Thus, the lawyers' view is that the key to claim avoidance is to get the contract right, and then to properly administer it. This, however, flies in the face of the causes of many construction claims and their disputes. These are:

- a lack of adequate communication between the parties which causes major difficulties, first in creating the situation from which the claim arises
- enabling it to be settled early and, therefore, more cheaply and without subsequent effect on time and attitude of the various construction parties.

So, will the legal framework prevent the development of a positive win/win approach?

The total (quality) management approach, beginning with the project management plan at the client inception stage, will ensure that the 'risk' allocation will be properly attributable.

The concept peculiar to both English and Australian law is that once parties make a contract, it is a bargain that the courts will not mend or remake for them if it goes wrong. This implies that

- the contract can be used to plan the operation
- the law will then assume this has been done and may demand exact and absolute performance of the bargain they struck
- the law will not recognize the 'doctrine of changed circumstances' to justify renegotiation or non-conformance.

This must, therefore, be done by the parties themselves, which requires pro-active management and leadership to develop a dynamic process of trust.

Partnering needs to build the human relationships to cope with the changed circumstances that is frequently the nature of construction contracts and so fill the gap between the initial planning framework at the time of contract and that which develops later.

Programming, brainstorming, benchmarking, budgeting, and evaluation techniques from the toolkit of TQM will all be needed to play their part but essentially within the human relationships developed by the partnering philosophy. Any construction contract should recognize the intended existence of 'partnering' just as it does the time, cost, and construction schedules contained within the 'contract documents'. Perhaps little amendment to existing standard forms appropriate to the works will be required in the UK construction industry.

Summary of Australasian experience

These studies are not of isolated examples of the approach. By February 1995 Thiess Contractors alone had completed 16 partnered projects and had four more in process for completion in 1995 and 1996. The first one (the Sunshine Motorway) was begun in September 1992 and finished in June 1993. These projects have a total value of (Australian dollars) A$275 million (£138 million) and cover both public and private clients. They range in cost from A$1.20 million to A$49.80 million (£570,000 to £23.7 million) with more than half below A$10 million (£4.73 million).

Thiess consider that the approach has been beneficial to many (but not all) of their projects. They have promoted open, frank communication, which in turn leads to further trust and the early resolution of issues. On some projects they have experienced a degree of scepticism and resistance to change particularly from certain clients. But the good results have encouraged them to actively promote partnering on *all* new Thiess projects.

In 1994 the winner of the Master Builders Award was the 120-bed five-level for six major clinical departments addition to the Nepean Hospital Project, New South Wales. Against a tight budget of A$24 million (£11 million) Barclay Mowlem Construction delivered the project on time, under budget and with no outstanding claims. It was the first partnering project undertaken in the public sector and the hospital remained fully operational throughout the contract. A further six hospital and other public sector projects have been commenced since with a value approaching £300 million.

The even more rapid growth of project partnering in Australia is reflected in the 12 nominations received for the Award in 1993 and the 18 in 1994, The Master Builders Association, which has in excess of 20,000 members, assesses that 300 projects, valued in excess of A$5–6 billion, already have, or are, embracing the partnering concept, which is not just confined to use by MBA members.

Approximately 30% of all non-residential projects are being partnered in Australia, and those contractors such as Theiss which have tried the concept and find it successful use it on all projects.

Public sector clients now have a policy of partnering a percentage of their projects. The NSW Public Works have a target of 20% for their major capital procurement projects, although their individual project managers are free to partner on any project. They believe that the cultural change necessary to embrace the concept is essential in the current highly competitive environment.

The view of Denis Wilson of the MBA, to whom I am indebted for the information relating to the MBA, is that consultants, including architects, have a sceptical attitude towards partnering which they think will remove some of their power, although this is probably more a myth than a reality as on the successful projects many of the accolades have gone to both consultants and sub-contractors.

New Zealand experience

In New Zealand too, 'partnering' has begun. The Museum of New Zealand, a NZ$150 million (£62 million) construction managed project for which Jasmax are the architects is being 'partnered' between the client and 16 contract package contractors—one of which is for the main building, one for the hard fitout and six for the building services.* The whole process is under the control of the construction project manager. Various partnering sessions have been facilitated by an expert. The success (or failure) depends, says one of those heavily involved, upon the personalities involved and a proper view of a truly shared goal – stated in the partnering charter. This was developed at a partnering workshop attended by the chief executives and top management of the client, the design team and the general contractors involved. As a consequence the current view is that more than 66% of the objectives are being achieved.

UK developments

A survey carried out in January 1995 of UK construction shows by comparison very little practical experience (see Appendix 2).

Outside of the petrochemical industry there exist few examples of the evolution of good project management with formal partnering arrangements, often these are not much more than negotiated design and build or less confrontational working between clients and contractors with previous joint project experience. However, some UK contractors, such as

*The author is indebted to John Sutherland, a Past President of the New Zealand Institution of Architects and a member of Jasmax, for this information.

the Morrison Group, have begun partnering on term contracts where over say 2–3 years an integration between client's staff and contractor's staff is taking place. Benefits for both are resulting in the administrative functions and in the speed of problem resolution and therefore cost.

John Laing have pioneered for a year or two 'down-stream' partnering with a preferred selection of sub-contractors and several other contractors in the UK are known to be following similar lines. But a full partnering project embracing client, design team, general contractor, and sub-contractors with a project charter and formal training and commitment sessions does not seem to have yet emerged.

Several commercial clients who have regular and rolling building programmes, such as supermarkets, petrol stations, retail distribution centres, are developing regular relationships with general and specialist contractors. Here, the incentive, with relatively simple construction projects, is generally to shorten the construction time and to build a regular continuing relationship, such as they experience in their main trading activity with their retail suppliers, from which general benefits will accrue to those involved.

One such initiative worthy of mention is that which developed on Safeway's Aldridge Store in the West Midlands with the Simons Group. This project, worth £5.8 million, was begun in 1992 with a 44 week construction programme. This was shortened after commencement on-site by 6 weeks to bring forward the opening date from February to before Christmas, and then, within 1 month of the revised opening date, shortened again by a further week – and achieved! This was a reduction of 16 % on an already tight programme.

Both Client and Contractor pooled their know-how through brainstorming sessions to achieve these considerable savings in time. The success of the operation led to further projects for Simons Group both as general contractors through a management contracting contract, and as designers on a fee contract, to be followed by a further and separate construction agreement on yet another project.

As yet, no formal 'project charter' and partnering agreements have been developed but the benefits to both stakeholders are clear and have been highlighted and aided by other activities common to project partnering such as the 'close-out' conference and project training sessions organized by the client. The approach is being carried forward into future projects of Safeways as client with other contractors and by the contractor with other similar clients. Simons Group assess they have started approximately 20 projects with Safeways and one other regular client with similar requirements and have carried the philosophy forward into over a hundred projects commenced since.

Simons Group's view is that whilst all sub-contractors on their projects are important the 'Pareto' 80/20 rule applies on each project and then some sub-contractors are more significant than others.

'Living in the real world' of competition and the ever-present constraints on management time both client and contractor are developing their approach to good management and improved performance empirically step by step and project by project. From discussions with both parties it is the author's view that even greater benefits would result to both, and their projects, from a further formalization of partnering through a charter. The full benefits result from the recognition that the full partnering technique applied both 'upstream' with the client and design team, and 'downstream' with contractors, results from the commitment that comes from the full awareness of the technique and common involvement in the creation of common objectives. All the experience in the USA and in Australia, and of partnering in the UK (outside of construction projects) supports this. Also, where a series of projects is involved the opportunity presents for 'Benchmarking' the results and using them as both a spur to, and guide for, the future, helping to maintain their leading edge as world-class organizations!

'Regular' clients such as supermarket chains are beginning to realize that the partnering concept developed in their core business can, and should be extended to their construction projects. They are actively working on the concept and the techniques necessary, working with approved suppliers for their mutual benefit.

Individual case study no. 1

Eielson Air Force Base, Alaska

Project	Transient Personnel Quarters Eielson Air Force Base, Alaska
General contractor	Osbourne Construction Company Fairbanks, Alaska
Size	$26 million public project
Stakeholders	33

The Transient Personnel Quarters was the first project in Alaska to be chosen as a partnering project between a contractor and the Corps of Engineers. The contract schedule required Osbourne Construction to complete 65% design documents and begin construction simultaneously. The Transient Personnel Quarters went from zero percent design to contract award in seven months. The total project from concept to finish took less than one and a half years.

The project was completed five months ahead of schedule even though 48 rooms were added to the original specifications. The project was further hampered by the discovery of hazardous materials, the small site size, and record breaking cold and snowfall (total accumulation of 152 inches).

Colonel Thomas O. Fleming, USAF Commander, said 'The level of co-operation between the Corps of Engineers, the contractors, and Eielson's civil engineers was magnificent. The results speak for themselves: moving from 0% design to contract award in seven months; bringing the project in 20 percent under programmed budget; exceeding an already compressed delivery schedule by five months; and perhaps most important, establishing a model for partnering in engineering construction'.

'Building the Eielson Transient Personnel Quarters using the partnering concept has been safe, efficient, claim-free and fun', said George Osbourne, president of Osbourne Construction Co. 'Partnering is a management system that gets people to attack problems, not each other. It is a way for owners, designers, and contractors to take charge of the construction process, and make it work to fulfil their mutual needs and goals'.

Complications arose when contaminated soil was found. The contractor hired a sub-contractor with portable laboratory. The Corps inspected the work, and the Air Force removed the contaminated soil, put it in drums, and moved it off-site.

In comparing this project to a similar project that was not partnered, the contractor found both projects had zero lost time accidents and quality construction. The partnered project however, has early project completion, profitability for all partners, effective communications, equitable negotiation, and timely dispute resolution, streamlined management and administrative processes, timely and quality design submittal and approvals, periodic and final evaluation of team performance, and fun.

To meet the challenge of a short construction season, Osbourne built a precast concrete plant, engineered the panels, produced a concrete mix design, and delivered the panels to the job site in less than 90 days. A total of 931 precast panels were produced and placed with 100% acceptance by the Corps of Engineers Quality Assurance Representative.

Many of the Transient Personnel Quarters stakeholders are now working on the Aircraft Weather Shelters, also on Eielson Air Force Base. A formal partnering process was established by a no cost change to the contract. This was a direct result of the success of the Transient Personnel Quarters. The Corp of Engineers Quality Assurance Representative and the Contractors Quality Control Manager developed a joint daily QA/QC report which is scheduled for review by a national joint committee of the COE and may be implemented nationwide. Trust, teamwork, and a general partnering spirit have carried over to the new project generating a positive environment and expectations.

Case study no. 2

USAA South-east Regional Office, Tampa, Florida

Project USAA South-east Regional Office

	Tampa, Florida
General contractor	HCB Contractors
	Tampa, Florida
Size	$60 million private project
Stakeholders	74

'Never in my professional career have I had as much fun on a project as I've had on the SERO project . . . and in producing the finest quality building that I've ever seen' said USAA's Project Manager Ron Roeder. Partnering was only natural for a company which had been named among the top ten of 'The 100 Best Companies to Work for in America'.

USAA directed the design and construction team project members to take a team approach to the construction process. HCB scheduled a two-day partnering and team building seminar with the contractors, architects, engineers, and owner. A partnering agreement was adopted and common goals and objectives identified. This partnering group met again mid-way through the project to assess progress.

A number of motivational activities were employed throughout the course of the project. After watching a quality orientation video and signing a commitment to quality certificate, each new employee (there were 1500) became a member the quality team. Even city building inspectors and officials signed the certificate as they came on the project for the first time.

The team concept was reinforced through monthly newsletters, monthly award luncheons, contractor/sub-contractor luncheons, and suggestion program. Team spirit permeated the project from the top on down. Rocky Sapp, who helped install the building's 12 elevators, commented, 'Normally, the quality of work goes unseen. It's nice to get recognized for the contribution. It helps to motivate people to do the best they can'. Field Engineer Jerry Albright agreed, 'I think that everyone who worked on the building feels a certain pride for it'. Even the people moving into the building noticed. 'It's the kind of building where you feel they didn't forget a thing', said a clerk-typist.

To reinforce the commitment, HCB Contractors placed full page ads in the Tampa *Tribune* which listed all the Quality Constructors by discipline. The purpose of the program was to do it right the first time and establish procedures aimed toward prevention, not correction. The program was spelled out in a booklet which outlined the goals, how to nominate someone for a quality construction award, how to submit suggestions (if a suggestion was accepted, the originator was allowed to leave one hour early with pay on Friday), safety program, designate responsibilities for earthwork, permits, environment protection, utilities, dewatering, layout, embeds, structure, roofing, etc.

USAA was not interested in value engineering. It wanted a quality building which would pay off in happy employees and good customer service. They did consider 'cost reduction suggestions'. For example, a wall in the lobby specified a particular wood veneer. HCS found through a sub-contractor that another material could provide the wood figuring desired at substantially lower cost. USAA accepted this suggestion. Another suggestion it accepted was the use of integral coloured stucco rather than normal stucco that would be painted later. This was a time and money saver for USAA

The quality initiative worked. Of more than 2000 embeds installed, only three required rework. HCB developed a project specific quality program which was updated, monitored, and statistically analysed over the course of the project. HCB will have the chance to use it again. USAA awarded HCB another major project as a result of the success of the SERO project.

Case study no. 3

Bonneville Navigation Lock, Cascade Locks, Oregon

Project	Bonneville Navigation Lock Cascade Locks, Oregon
General contractor	Kiewit/Al Johnson Vancouver, Washington
Size	$140 million public project
Stakeholders	9

When it had to replace a 50-year-old undersized lock at Bonneville Dam on the Columbia River near Portland, the Portland District Corps of Engineers, then commanded by Colonel Charles Cowan, knew partnering was the only way to accomplish this mammoth project.

Work on the Bonneville Navigation lock began in 1986 and opened to traffic in March 1993. The site was congested. The existing navigation lock was kept in operation while the new one was constructed. Also at the site were two operating power houses which supplied much of the north-west's electricity; a railroad; a visitor's centre which attracted 400,000 visitors per year; a popular fishing hole; salmon hatchery; and active slide area. In order to protect the salmon run, no in-water work was permitted from mid-November to March 1.

To get started, the Kiewit/Al Johnson Project Manager and the Corps Resident Engineers spent a week together at a leadership school getting to know each other. About 40 people from the Portland District Corps of Engineers, Kiewit/Al Johnson, and some of the major sub-contractors, attended a three-day workshop to learn partnering skills. They developed goals for safety, quality, schedule, issue resolution, maintaining integrity of public areas and fish hatchery, value engineering, cost growth, minimizing paper work, preventing litigation, and making the project enjoyable.

When a new resident engineer was appointed a few months into the job, he too spent a week with the project manager and participated in evaluation sessions.

An issue resolution procedure was developed which stated:

- no matter who owned the problem, it was everyone's task to solve the problem at the net lowest cost and the net lowest impact to the job schedule
- no action on a problem was not an option
- issues were solved at the lowest possible level. If the problem could not be solved at that level, it would be escalated to the next level until resolved.

A number of recognition programs kept employees motivated. They included safety incentive plans, outstanding partnering awards to individuals, certificates of appreciation, commendations letters, monetary awards, staff lunches, and picnics. Employees cruised on the sternwheeler *Columbia George* to commemorate the first passage through the newly-completed lock and the dedication ceremony included speakers from each member of the partnership. The job was shut down one Saturday midway through the project so families could tour the job and join in a picnic. Every two weeks, the project manager and the resident engineer walked the entire project and awarded the area of best or most improved quality the 'quality flag' to be flown over their work for the next two weeks.

Value Engineering Joint Proposals workshops were held early in the project. Eliminating an overpass of the haul road saved $600,000 and changes to the cross-sectional shape of the downstream guidewall saved another $600,000. Thirty-nine other ideas were implemented with savings of over $4 million.

The Bonneville contract required nearly 700 lift drawings detailing the structural concrete and estimated at 350 effort weeks to complete. In an early partnering session, it was suggested that if those drawings could be manipulated, little new drawing would be required. The Corps drawings were downloaded into the contractor's CAD system. The access to these drawings nearly doubled the lift drawing production.

In order to meet the March 1993 deadline for opening the lock, it was crucial to complete the punchlist. The punchlist was developed jointly by the prime contractor, the sub-contractors, and the owner. The team met weekly to assign priorities and set goals. By sharing notes, field sketches, and files, the team was able to negotiate priorities, identify remaining work issues, and obtain fair settlements.

'Partnering is an exciting management technique in construction', said David Brown, project manager for the lock, 'partnering has made the daily work between the contractors and the Corps much easier because we talk to each other rather then sending letters'.

Case study no. 4

Chevron Wastewater Treatment Project, Port Arthur, Texas

Project	Chevron Wastewater Treatment Project Port Arthur, Texas
General contractor	H.B. Zachry Company San Antonio, Texas
Size	$100 million private project
Stakeholders	5

When a progressive owner like Chevron Products Company and a quality constructor like H. B. Zachry team up, the result is a job well done. Chevron's perspective of the project was somewhat unique; its highest priority was creating a successful project team, not hiring the lowest bidder. The paradigm shift worked because the Chevron Wastewater Treatment Project ran up an impressive list of statistics:

- it was completed two months ahead of schedule
- there were no lost time accidents with an incident rate of 1.1 in 1,792,000 effort hours
- the project came in 10% under budget
- there was less than 2% rework.

Chevron and H. B. Zachry worked together to build teamwork and enthusiasm. Employees were treated to coffee and donuts while they waited for buses to take them to the job site. They were also offered aerobic exercises. A variety of motivational activities were employed throughout the project, including:

- 'Town Hall' meetings with construction and design personnel
- personnel and group recognition
- milestone lunches for design and construction personnel
- safety picnic for the whole project
- belt buckle design and award
- weekly safety drawings
- recognition for excellent performance
- personnel recognition letters
- hard hat stickers.

Chevron initiated formal team building activities to promote communications and provide a dispute resolution mechanism. An outside facilitator trained the owner, engineer, general contractor, and departments and crafts within these organizations. Participants developed expectation and buy-in workshops. Quarterly meetings with the owner, engineer, and general contractor covered team performance, plant commissioning and start-up, plant operation, appropriations request budget, and schedule.

Areas of concern were identified and appropriate action taken. 'Lessons Learned' workshops were held at 70% project completion, which further strengthened the teams and sped the project to its conclusion.

Management teams were created and empowered to resolve disputes at the job site level. Corporate sponsors participated on these teams. Subcontractors were involved in the workshops, milestone celebrations, and recognitions. Craft involvement was taken to the foreman level where expectations were discussed, agreed upon, and written down for all to see and understand. Communication and commitment were heard throughout.

Early involvement of construction personnel during the design development resulted in savings of over $7.5 million. Process, or peer, reviews were valuable in uncovering simple design flaws.

The reward for a job well done? An additional $7 million for a raw water clarifier and storm drainage improvement was added to the original contract and H. B. Zachry Company received another contract for $30 million in the same plant with the same team.

Case study no. 5

The Lacey V. Murrow Replacement Bridge, Seattle, Washington

Project	The Lacey V. Murrow Replacement Bridge Seattle, Washington
General contractor	General Construction Company in joint venture with Rainier Seattle, Washington
Size	$88 million public project
Stakeholders	32

What has 67,000 cubic yards of concrete, 32,500 feet of anchor cable, over 1,000,000 feet of electrical wire, and 13,500 tons of reinforcing steel? What involved 170 change orders with no claims?

It is the Lacey V. Murrow Floating Bridge Replacement. The LVM Bridge was a high profile project, given that two other bridges had sunk in the last 15 years. In order to produce a quality product and speed over 110,000 motorists each day on their way, General Construction Company in a joint venture with Rainier and the Washington State Department of Transportation implemented partnering and completed the project one year ahead of schedule.

In order to make certain the partnering effort did not falter, team members met formally each quarter to evaluate progress. A comment/rating sheet was filled out ahead of each meeting in order to monitor progress and identify areas for improvement. WSDOT's headquarter's design personnel participated in bimonthly meetings and job site tours, enabling them to develop personal relationships with other team members and witness issues being resolved between team members. The project began with a two-day partnering training session as the group set goals, defined objectives, identified potential problems, and developed an issue resolution system. Letter wars were eliminated when the partners agreed that there would be no bad news letters without prior discussion. WSDOT agreed to complete submittal review and respond within 14 days. In the spirit of partnering, they replied on average in seven days.

The partnering skills came in handy in keeping just the right amount of a mineral admixture on hand. The supplier could only store enough for one day's production and the supply source was seven to ten days away. The sub-contractor attended meetings and was able to meet the delivery schedule by being aware of issues before they became conflicts.

James Repman, President of Lone Star Northwest, Inc, the sub-contractor, had this to say, 'this spirit of communication carried over into all areas of the project in which we were involved. Whenever there was an issue regarding concrete, we were invited to attend meetings and to have meaningful input. Issues such as form pressures, concrete slumps, colour differences, etc., were all resolved before they became sources of conflict. This project is the first we've been involved in that regularly held meetings just to see if there was anything that could be done better. An example of what derived from the 'think' sessions included forming the anchors on a barge then floating them to our dock to be poured. We believed strongly enough in the people and the process that we recently selected General Construction to rebuild our cement dock in Portland. . . and suggested a partnering form of contract for the project . . . '.

A naval architect had to review and approve certain construction plans. In order for the naval architect to keep pace with a demanding and constantly changing schedule, a computer model was modified to analyse the pontoons during construction and simulate the impact of planned events before performing the work.

Many innovative techniques were used on the project to expedite construction:

- interior walls and portions of anchor galleries were precast
- a craft incentive program encouraged safety
- bridge anchors were formed, poured and stripped on a barge, which eliminated crane and ready mix transport expenses
- underwater cameras assisted with anchor placement, eliminating the need for a diver to guide the 300 ton blocks of concrete.

With field tolerances of ½ inch, rebar restrictive walls, 67,000 cubic yards of concrete, over 75,000 cast-in-place inserts, only minor rework was required. This was accomplished through detailing over 2,400 field drawings, erecting test sections to prove and improve construction techniques, and modifying techniques throughout the project's duration to provide a better quality project.

Thomas Nelson, Chief Construction Engineer for the Washington State Department of Transportation commended the extraordinary effort and project management know-how of General Construction Company. 'There were many problems on this project that could easily have resulted in an adversarial relationship between WSDOT and General Construction. In all cases the field and administrative staff demonstrated an attitude that minimized conflict and focused on solving the problem cooperatively. This project could have been very stressful, but turned out to be a very challenging, yet positive experience . . . . We look forward to working with you on future projects'.

Case study no. 6

Menasco Aerosystems Manufacturing Plant Expansion, Fort Worth, Texas

Project	Menasco Aerosystems Manufacturing Plant Expansion, Fort Worth, Texas
General contractor	r-o-s Constructors, Inc. Fort Worth, Texas
Size	$3.8 million private project
Stakeholders	30

The Menasco Aerosystems was closing a facility in another state, so this existing facility needed to be expanded to accept extremely large metal working equipment from the closing facility. The expansion had to be designed, built, and substantially completed in 185 days.

A decision was made to design-build in order to meet the schedule and budget. Partnering became one of the most valuable tools to the entire project.

An initial partnering session was held at the beginning of the project with the owner, general contractor, architect/engineer, and mechanical and electrical sub-contractors. They set measurable 'SMART goals', which reflected the mission statement. Following the initial partnering session, a Quality Steering Committee was formed to oversee evaluations and administer recommendations. This rotating member committee met every two or three weeks to evaluate the teams' progress. Discussions centred on positive suggestions and ways to address problems and meet goals. Follow-up sessions were held and sub-contractors welcomed to the team as they came onto the job. The team charter also included teamwork and communications objectives and a conflict resolution system.

Milestones were important in this project because of the fast-track completion, so critical path events were reviewed at weekly coordination meetings. When the structural frame was completed on schedule, there was a barbecue for all team members, including the field personnel. Key team members received Certificates of Appreciation for their extraordinary efforts in meeting such an aggressive goal.

The architect and general contractor evaluated several scheduling options using computer software. The same software was used by sub-contractors to create detailed mini-sub schedules for daily coordination with other trades at the job site. This proved valuable in resolving daily scheduling conflicts. A written issue escalation process identified personnel and time lines for resolving issues within one day.

A team of owner personnel, designers, and the contractor developed a scope of work for the project. The original scope of work exceeded budget so through an honest exchange of ideas, the team prepared a list of cost-saving ideas, including foundation redesign and revising the chilled water cooling system to rooftop HVAC units. Costs were cut by 25% and met owner satisfaction.

A large portion of the construction budget was allocated to the mechanical and electrical systems. Because of their importance in the design and completion of the project, the mechanical and electrical sub-contractors were offered an up-front percentage in project savings. All major sub-contractors and suppliers were involved in at least one formal partnering session. Preconstruction meetings for sub-contractor job crews were held so they could review quality requirements. They also participated in weekly coordination meetings in which the mission statement and SMART goals were reviewed. Through these meetings, many problems were resolved by team members working together.

The Menasco Aerosystems Plant Expansion was the first exposure to partnering for most of the team members. Beginning the project, the barriers of schedule, design, and budget constraints seemed insurmountable. After the initial partnering session, the team emerged with an attitude of commitment, plans for action, and mutual respect and cooperation.

Brad Welhouse, Vice President of NOW Construction Inc., says of partnering, 'use it and use it often. For a small sub-contractor such as ourselves, we keep a limited staff and try to run lean and mean. Taking on a job of this magnitude is a real test of will, wit, and patience, but knowing that the other trades we are working with are cooperative and ready to help made us feel comfortable and team-worthy. Having the staff at r-o-s constantly remind us of the tasks at hand allowed us to stay on top of areas of our work. Fast track projects often produce mistakes which then release personal pressure as a by-product. Knowing we were on the team made the unexpected situations workable solutions'.

Case study no. 7

Mayflower Homes Health Center, Grinnell, Iowa

Project	Mayflower Homes Health Center Grinnell, Iowa
General contractor	Story Construction Co. Ames, Iowa
Size	$3.8 million private project
Stakeholders	39

At first glance, the Mayflower Homes Health Center doesn't look like a nursing home. The design and quality of the home makes it look just like that – a home. Partnering made it happen.

Ted Mokricky, Executive Director of Mayflower Homes Inc., says it best, 'I wasn't sure that partnering would make a measurable contribution to our project. But simply stated, the end result was that the project was completed on schedule and under budget. I think there were many times that I forgot Dan Ward, the project superintendent, was an employee of Story Construction. I came to think of Dan as one of the Mayflower staff, and this is true of some of the other key subs too – the electrician, mechanical people. They really presented themselves as part of the Mayflower team. Relationships were enhanced because of the partnering process. The number of change orders were reduced. Problems did appear but they were resolved at the lowest level possible. Many times I was not involved in the problem identification and solution process. Problems were resolved by some of my staff and by the subs and by the Story personnel before they even came to my attention, which I appreciated very much. To wrap it up, on our next project we certainly will adopt the partnering process again, and I would expect partnering and would make that a requirement for any future project that Mayflower is involved in'.

Dan Ward responds, 'the real measurement of success was evidenced almost every day throughout the project. A lot of times problems would arise and within the hour solutions were analysed and hashed over and decided upon and we were off building a health care centre again. When we needed input from higher levels, we approached those people with solutions instead of problems. That was probably the most successful thing that I saw come out of partnering because it kept the construction going. No time was spent trying to build a case for yourself or find fault with others. We set out to create a home, not an institution, and I believe we were successful at that. One of our goals was to have fun. We had problems, but our time was spent solving the problem so we could start having fun again. We had a lot of fun and a lot of good relationships were built'.

To implement the partnering process, an evening social event was held for participants to get to know each other on a first name basis. Story's project superintendent hosted luncheons off-site after each monthly progress meeting for sub-contractor foremen and other key employees. A wrap-up meeting was held that included golf, workshop, and dinner. The session was a celebration of the completion of the project and as motivation to get punch lists completed and promote use of partnering on the next project. It worked, because Mayflower awarded a contract to Story on a negotiated basis for 21 independent living units. Previous construction projects had been awarded on a competitive bid basis.

Value engineering ran rampant on this project. Team members worked together to:

- generate ideas to meet fire code
- identify and find space for HVAC ductwork
- substitute sand backfill for specified dirt backfill in some locations
- review, expedite and prioritize structural steel shop drawings (contractor, designer, and steel supplier)
- find cost savings on drywall
- improve EIFS system which saved money and provided better warranty
- coordinate construction of skywall
- improve sound/weather/telephone system (electrical contractor)
- develop system to expedite and authorize change orders prior to distribution of paperwork
- streamline payment estimate approval process for timely payments to contractors.

The project came in under budget, ahead of schedule, with no claims.

Case study no. 8

Morris Sheppard Dam Stilling Basin Possum Kingdom Lake, Texas

Project	Morris Sheppard Dam Stilling Basin Possum Kingdom Lake, Texas
General contractor	Martin K. Eby Construction Company Inc. Bedford, Texas
Size	$3.5 million public project
Stakeholders	4

There were several potential problems confronting the Morris Sheppard Dam Stilling Basin project. At the partnering meeting, the team identified 30 potential problems, discussed them, and then developed actions. The participants believe that partnering gave this project the boost necessary to be completed in a quality and timely manner. The charter provided a series of common and mutual goals for the entire project team. When changes occurred, the team focused on finding workable solutions. The problem solving techniques initiated during the workshop were successfully used.

Placement of 9000 cubic yards of roller compacted concrete on the south basin of the Brazos river presented a unique challenge. Not only was access limited, but the concrete was to be placed on a steep incline in one foot lifts. An employee involvement team discussed possible solutions, and finally decided to use a dozer box with a spectra-physics laserplane attached to the blade. The laser attachment indicated proper placement of the lifts, and saved engineering and placement time.

The contractor performed some site and road improvements to a site in return for using the site as a needed source of clay for the cofferdam construction.

Constant attention was given to scheduling, safety, quality, and co-ordination of activities. Frequent exchange of ideas for improvement occurred between owner, engineer, contractor, and sub-contractors. It was important to continue development of on-going relationships among team members. The initial partnering workshop focused on personal work styles, individual characteristics, and team building. Team members were able to voice their concerns, develop common and mutual goals, and work together to meet those goals.

John Jarboe, Reservoir System Construction maintenance manager for the Brazos River Authority, said, 'I sure had fun along the way and hated to see the job end. I believe the partnering conference set the tone for the project. The team was put together and everyone came away knowing the players, their authorities, responsibilities, and concerns. I think this allowed us to support each other and achieve [our goals]. When placing excavated material in the water for cofferdam construction caused a lot of turbidity, the work plan was changed to use washed rock available on-site. This was extra work for the contractor but supported our environmental concerns expressed at the conference and incorporated [one of the goals]. We were keenly aware of maintaining the contractor's schedule to accomplish cofferdam removal before December 1'.

He continues, 'much of the success of the project can be attributed to the good communications between the team members. Many of our problems were worked out in on-site discussions. We all developed trust for one another. Our word was as good as a written document. It was not necessary to spend our time preparing documentation of decisions made to put in the files. We could pursue getting on with the work. As a retired Corps of Engineers employee, I still remember how it was 30 years ago. We drew a line in the dirt at the pre-work conference and lined up on opposite sides as opponents. We were convinced that it was our job to force the contractor to do the job as planned and produce a quality product. We thought he was only interested in getting done and extracting a large profit. We learned early in the job that our contractors did totally support our goals and this made it easy for us to support the contractor's goals'.

The team achieved all goals outlined on the charter. The project was completed 13 days ahead of the original schedule and four months ahead of the contract completion date. There were no change orders, no claims, no punch list, and costs were kept in alignment with Brazos River Authority contingencies.

Case study no. 9

North Dandeloup Dam Project

This A$66 million (£30 million) project South East of Perth in Western Australia for the Western Australia Water Authority, was let to McMahon Construction on a 'value for money' basis in August 1992 on a Schedule of Rates contract. After tenders in March 1992 from seven pre-registered contractors, it commenced in October 1992 and was completed in June 1994 and came on stream in December 1994.

From an initial two-day partnering workshop, which approximately 40 people attended in equal numbers from client and contractor, it developed:

- a project charter
- an escalation (ADR) process
- an evaluation process
- an action plan.

The critical success factors were found to be as follows.

Commitment

Demonstrated commitment from the highest level of both organizations. In the first six months the Chairman and the CEOs of both organizations visited the site twice – together. Senior executive and senior operations staff visited at least fortnightly, and met regularly.

Effort

Time and effort are required to make partnering work – far more than required on contracts without partnering, especially by senior and supervisory staff. Partnering creates the environment and provides the processes – the effort comes from those involved in the contract at all levels.

Against these critical success factors are the following.

Reduced paperwork

The paperwork on this contract is approximately half that of previous contracts. All the contractual requirements for documentation are being met. With the need for case building minimized, the nature and volume of documentation is considerably reduced.

Trust

The development of trust is the cornerstone of the partnering concept. It starts with the workshop – but every opportunity should be taken to develop a relationship of trust between all parties – at all levels.

This is the 'warm, fuzzy' part of partnering which contractors and owners alike felt uncomfortable with. It has not been a dominant part of construction contracting in Australia if indeed it existed at all. The 'gotcha' principle is usually alive and well in both camps.

Effort required

The problems which occur on a large contract remain – it is only the manner of resolving the problem which changes. Resolution is not any easier. It required a great deal of effort and dedicated work together to resolve the problems. With partnering as a basis (trust, commitment, equity, etc.), common, acceptable resolutions have come about.

Case study no. 10

Science and Technology Building, Macquarie University

Project size was approximately A$12.77 million = £6.4 million.
Let on amended A$ 2124 for a 46-week construction period.
Partnering sponsored by the Gyles Royal Commission.
Architects: Devine Erby Mazlin. Contractor: Leighton Contractors

The building was a comparatively simple design and not fast-tracked

The state of the documentation of the project resulted in over 240 architects' instructions but primarily, these related to variations of building services initiated by the client. Taking advantage of low prices Macquarie University increased the complexity of the services within the building so that ultimately the services accounted for 40% of the contract value.

Documentation for these variations was late in production and prompted a delay claimed by Leightons, which was settled without litigation or arbitration.

Educating the parties about the concept

Apparently the only attempt at doing this was to give each of the participants a parcel of documents on partnering prepared by Cowan's original employers, the US Corps of Army Engineers. This lack of education about the concept seems to have perpetuated adversarial relationships during initial partnering sessions. Improved teamwork and more open lines of communications between the participants were a late development.

No attempt seems to have been made to choose participants from participating organizations who were sympathetic to the concept.

Macquarie University had limited experience in large construction projects and in the traditional way relied on the architect as superintendent. This created communication problems. It prevented Leightons effectively communicating with the client save through the upper levels of the issue resolution procedure, something that became very important when Leightons prepared and brought their delay claim.

Making the commitment

No partnering leaders were chosen and it seems that there was a lack of top management support within Macquarie University and Leightons. On the client's side, the University Vice Chancellor took no part in partnering sessions and Leightons' nominated off-site senior management representative attended only two out of the seven partnering sessions.

The partnering workshop

There was no partnering workshop. As a result participants did not have the opportunity to analyse potential problems in the course of the contract, the 'rocks on the road' or 'boulders in the bush' as they are often referred to in Australia, or to get to know each other or to develop mutually agreed objectives for the project. The Royal Commission assigned a facilitator and he prepared a partnering charter and implementation and monitoring plan. He prepared this independently of the participants although it was later endorsed by them.

Implementation of the joint evaluation process and final evaluation

Partnering sessions were held once every six weeks. There was markedly better attendance from on-site representatives than off-site representatives. Only a limited number of the 57 sub-contractors on the project participated in the partnering sessions.

The partnering charter's objectives

The partnering charter had ten objectives. No doubt some of the difficulties experienced in partnering the project were attributable to the lack of ownership of this charter by the project participants.

The ten objectives

(*a*) Commitment to a quality project by:

- providing quality documentation and meeting the design intent
- devising a joint workmanship and materials quality management programme
- building it right the first time

- exploring and utilizing new technologies and methods for technical excellence.

(*b*) In safety performance by completing the project with the following results:

- no fatalities
- no lost time injuries
- no general public liability over A$500.00 (£230).

(*c*) Commitment to on time opening of the building for its intended use by:

- timely resolution of disputes
- joint management of the schedule
- maintaining a steady, uniform work flow.

(*d*) Commitment to containment of the cost of the project by:

- limiting total cost growth to less than 5%
- minimizing contractor and sub-contractor costs
- providing ample warning of variations, deletions and additions
- providing speedy design and costing of variations, deletions and additions
- minimizing paperwork and red tape
- making quick and joint decisions after joint consultation
- encouraging information sharing at all levels.

(*e*) Commitment to creating a high level of industrial relations harmony by:

- regular meetings with the workforce to discuss technicalities and programme
- pre-emptive attention to the quality and standard of amenities
- prompt settlement of disputes and answers to questions on statutory rights
- minimizing incidence of lost time, paid or unpaid
- strict compliance with the law and agreed procedures by all.

(*f*) Achieving contract completion without the need for contractual arbitration or litigation.
(*g*) Achieving real teamwork at site level by fully involving all participants in the process particularly sub-contractors and their employees.
(*h*) Administering the contract so that the client, the consultants, the contractor, the sub-contractors and workers are treated fairly.
(*i*) Minimizing disruption on the campus during construction by:

- communication by contractor of deliveries and construction activities

- communication by client for specific campus activities and programmes.

(*j*) Making the project enjoyable through partnering at all levels

Limited achievement of objectives

The full satisfaction of all of these objectives was limited, in particular the following.

(*a*) Objective 3, timely completion, was affected seriously by the client's variations to the building services. Of approximately 55 days delay, 30 days were attributable to these variations.
(*b*) Objective 4, containing the cost of the project to less than 5%, ignored the potential for increased costs with the variations and the attendant disruption or delay. Total project costs, excluding the delay claim, came in at about 6% above the original contract value and at about 7% if the delay claim is taken into account.
(*c*) Objective 5, a new high level of industrial relations harmony seems to have been achieved. There were no significant industrial relations problems and minimal lost time.
(*d*) Objective 6, completion of the project without arbitration or litigation – it was achieved but only just. The presentation of Leightons' delay claim tested this objective.

An issue resolution process had been established as part of the partnering implementation and monitoring plan by the Royal Commission independently of the participants. Essentially it provided that any dispute at the level of on-site management which had not been settled within two weeks would be drawn to the attention of an off-site tier of management who would then have two weeks to settle the dispute. If they were unsuccessful, the dispute would immediately pass to the chief executive/senior party of each party for resolution.

Ultimately Macquarie University and Leightons settled the delay claim prior to the last partnering session. It was settled, however, at the level of off-site senior management not at on-site level.

(*e*) Objective 7 was to achieve a fuller participation by participants including sub-contractors and their employees. However, only a limited number of the 57 sub-contractors participated in the partnering sessions.

Author's conclusions

The deficiencies in the process probably result from:

- the lack of preparation and education on the concept reducing the potential for success
- insufficient top management support

- the absence of a partnering workshop hindered teamwork, trust and the development of a single management team with shared objectives.

No doubt other criticisms should be made. At the end of the day, however, partnering seems to have kept the parties talking, contained unjustifiable cost and prevented arbitration or litigation of the delay claim cost.

Perhaps the greatest error was the 'imposition' of the concept on the parties who therefore did not truly 'own' the process, rather than providing an extended education process of the top management of *all* stakeholders – client/design team/contractors and the key sub-contractors in a carefully structured workshop at which they could work out the project charter for themselves and so create a new and very different tailor-made *project* culture.

Case study no. 11 (building)

Green Island Redevelopment, Cairns, Queensland, Australia

Client	Green Island Resort Project
Architect	DBI
Contractor	Thiess Contractors

The project

Redevelopment of an unspoiled tropical island on the Great Barrier Reef about 16 miles off Cairns in North Queensland. It has a unique natural environment and spectacular flora and fauna. The work was undertaken in two contracts. The first, a lump sum fixed price for support infrastructure, carried out between April and November 1992. This was a preliminary to a two-stage resort facility.

(*a*) Stage 1. A 'Day Tripper' reception building with a swimming pool, restaurant and staff accommodation which will later serve Stage 2.

(*b*) Stage 2. Five-star hotel suites in 46 low rise units of one- and two-storey construction in rendered blockwork with shingle roof.

All buildings are suspended 1 metre above ground level to protect vegetation and minimize root damage. All are connected by timber decks and walkways.

The original contractual strategy allowed the civil works to proceed while the design of Stages 1 and 2 was developed but the redevelopment as a whole was subsequently put on hold with the result that the design was not complete before the civil works were completed in November 1992.

The work within Stages 1 and 2 was a negotiated construction management contract with a fixed lump sum for preliminaries and overheads and managed trade sub-contracts. The trade contract was entered into on a cost reimbursement basis with the head contractor being paid a negotiated management fee.

Work started on the second contract Stages 1 and 2 in January 1993 when the design documentation was not complete. The structural engineers were based in Brisbane about 1000 kilometres south of the project, the principal, the operator and the head contractor and the services engineers were based in Cairns, the civil engineer was based in Townsville a city about 500 kilometres from the project, and the architect was based on the Gold Coast further to the south of the project than Brisbane. This gives some idea of the logistical nightmare that could have resulted if it had not been possible to identify and resolve problems at a very early stage.

Demolition and site clearance in January 1993 was followed by 'fast track' construction. The term 'fast track' refers to the development of design concurrently with construction. Piling started shortly before March. Structures were founded on small driven pre-cast concrete piles set about 1 m above ground.

Partnering conference

It was the first time the principal had partnered a contract, and the partnering conference was held in March 1993.

The partnering conference, which was assisted by an outside consultant as 'facilitator', produced the partnering charter whose objectives are set out below.

Partnering objectives

'We are committed to implementing the partnering process throughout this project, with the following objectives:

- provide interesting, enjoyable and satisfying work for all
- achieve or better program targets
- carry out a safe job
- maintain an effective ongoing team relationship
- generate future work
- carry out appropriate, timely and effective verbal and written communication
- resolve all issues at the lowest level in a timely and progressive way
- build a quality job
- achieve community and environmental expectations
- minimize the impact on construction and tourist operations
- achieve all parties' financial goals'.

Note the mixed performance and relationship objectives. Note too that one of the objectives was to generate future work.

There were no sub-contractors at the partnering conference. Partnering was between the principal, the clients and the consultants. The performance of the sub-contractors was thus the head contractor's responsibility and major sub-contractors were appraised of the partnering arrangement individually. Generally the arrangement was well received by sub-contractors.

The issue resolution form (see Fig. 33) produced by the partnering conference may appear complex because of the involvement of the consultants but it appears to have forced the resolution of problems before they interfered with job progress.

Managing partnering

There were a few healthy sceptics about the partnering process but everyone persisted with it. There were project control group (PCG) meetings initially almost weekly and a full PCG meeting once a month. It was prior to the full PCG meeting that the partnering assessment forms were filled out by someone on behalf of the principal. A partnering assessment summary was also completed (see Figs 34 and 35).

Assessment forms were completed by the contractor, the principal and consultants, and each one of those was to be studied before a PCG meeting and the score for the individual against the average for that objective noted to facilitate consideration of individual performance against the objective in the PCG meeting.

A representative of the principal and a representative of the head contractor took it in turn to chair the partnering sessions in the PCG meetings and go through the results. Each discussion like this produced a partnering report and action plan (see Fig. 36).

Conclusion – what partnering achieved at Green Island

The incompleteness of the design documentation and difficulties in documentation dissemination with the parties in so many different locations could, under another contractual scheme, or without partnering, have produced a quite disastrous project. As it was, difficulties were flushed out and dealt with promptly. Each session produced the report and action plan for the next month after discussion of the scores achieved and each subsequent PCG produced a review of the action plan and another action plan. What this meant in terms of project success was completion of Stage 1, the day trippers' facility, in July 1993, six weeks ahead of time, and without payment of acceleration costs.

There was a great reduction in correspondence even if there was an increase in meeting time. The fact that partnering is more time consuming has to be recognized, but it must also be noted that the time is spent not in posturing or defensive correspondence but in discussion in a blame-free atmosphere allowing formulation of pro-active procedures for

ANNEXURE 7

DAIKYO

GREAT ADVENTURES

GREEN ISLAND RESORT PROJECT
CAIRNS • AUSTRALIA

THIESS

ISSUE RESOLUTION PROCEDURE

LEVEL	THIESS	DAIKYO	GREAT ADVENTURES	EQUA	DBI	MCWILLIAM & PARTNERS	GHD	LINCOLN SCOTT	RIDER HUNT	TIME FRAME
1 SITE (Non cost/ time).	Site Super. Don Gordon	Dave Webb (Landscape)	Island Mgr Peter Gleeson Engineer Alex Taylor		Site Architect Keith Whenmouth	Mr "X"	R Blain (Civil) R Buff (M&E)	Darren Bilsborough		1 hour
2	Don Gordon Contract Adm Rick Thompson Proj. Eng John McHugh	Proj Mgr. Glen Busby		Lloyd Groves	Keith Whenmouth Dave Brown	Peter Melloy	John Gersekowski	Darren Bilsborough (Mech) Phil Dorne (Elec)	Ken Spain	1 day
3	Henry Gartside	Glenn Busby		Charles Hart	Dave Brown				Ken Spain	1 day
4	Bill McClanachan	Doug Lambert	Reg Reddicliffe						Peter Spencer	2 weeks

GROUND RULES:

1. Resolve issue at lowest level
2. Unresolved issues escalated upward by both partners prior to causing project delays or costs.
3. No jumping levels of authority.
4. Ignoring the issue or "no decision" is not acceptable.
5. Time frame may be extended by mutual agreement, for data gathering.

Figure 33. Green Island issue resolution procedure

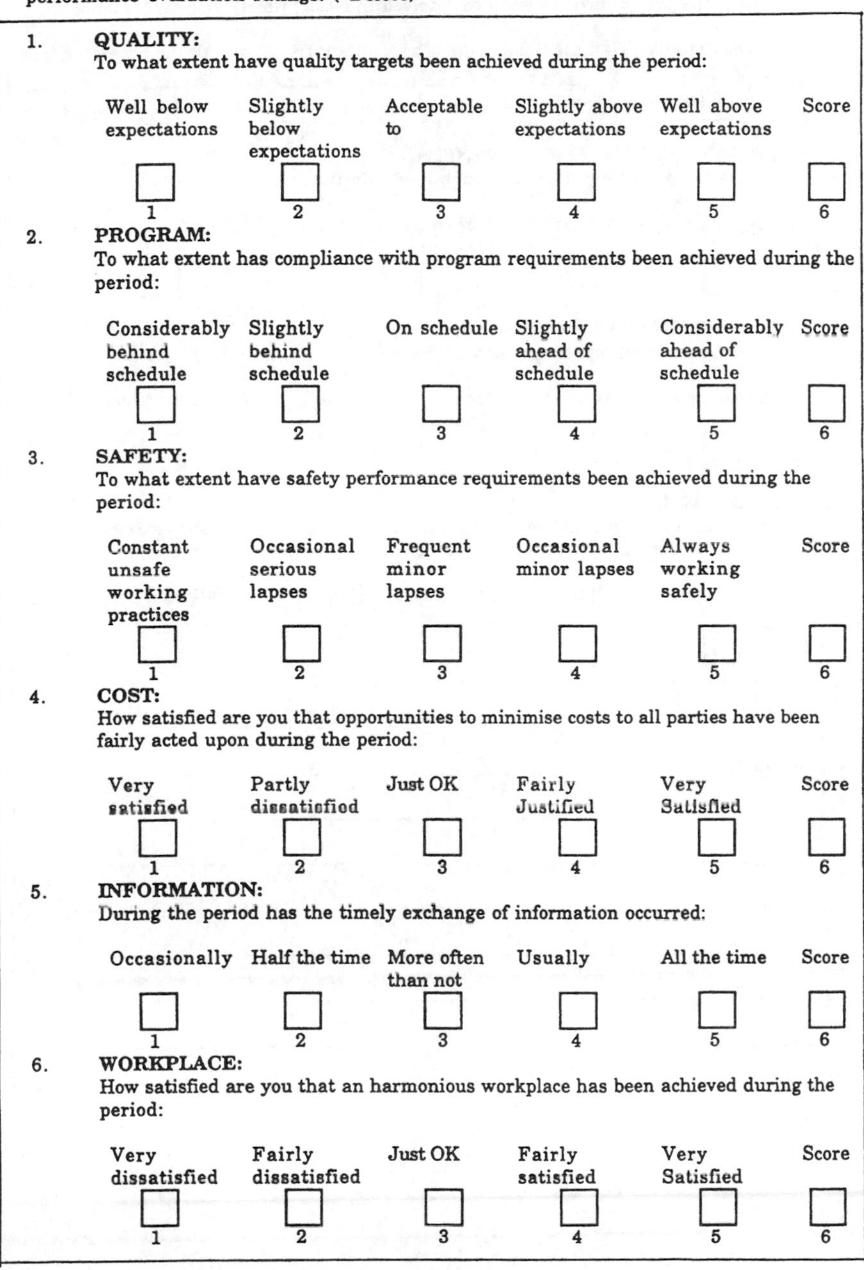

Page 1 of 2

PARTNERING ASSESSMENT FORM

Over the past week/fortnight/month* the partnering team has achieved the following performance evaluation ratings: (*Delete)

1. **QUALITY:**
To what extent have quality targets been achieved during the period:

Well below expectations	Slightly below expectations	Acceptable to	Slightly above expectations	Well above expectations	Score
☐ 1	☐ 2	☐ 3	☐ 4	☐ 5	☐ 6

2. **PROGRAM:**
To what extent has compliance with program requirements been achieved during the period:

Considerably behind schedule	Slightly behind schedule	On schedule	Slightly ahead of schedule	Considerably ahead of schedule	Score
☐ 1	☐ 2	☐ 3	☐ 4	☐ 5	☐ 6

3. **SAFETY:**
To what extent have safety performance requirements been achieved during the period:

Constant unsafe working practices	Occasional serious lapses	Frequent minor lapses	Occasional minor lapses	Always working safely	Score
☐ 1	☐ 2	☐ 3	☐ 4	☐ 5	☐ 6

4. **COST:**
How satisfied are you that opportunities to minimise costs to all parties have been fairly acted upon during the period:

Very satisfied	Partly dissatisfied	Just OK	Fairly Justified	Very Satisfied	Score
☐ 1	☐ 2	☐ 3	☐ 4	☐ 5	☐ 6

5. **INFORMATION:**
During the period has the timely exchange of information occurred:

Occasionally	Half the time	More often than not	Usually	All the time	Score
☐ 1	☐ 2	☐ 3	☐ 4	☐ 5	☐ 6

6. **WORKPLACE:**
How satisfied are you that an harmonious workplace has been achieved during the period:

Very dissatisfied	Fairly dissatisfied	Just OK	Fairly satisfied	Very Satisfied	Score
☐ 1	☐ 2	☐ 3	☐ 4	☐ 5	☐ 6

Figure 34. Green Island partnering assessment form (*continued overleaf*)

PARTNERING ASSESSMENT FORM

7. **RESOLUTION OF ISSUES:**
During the period have issues been resolved in a fair, timely and early manner:

Occasionally	Half the time	More often than not	Usually	All the time	Score
☐ 1	☐ 2	☐ 3	☐ 4	☐ 5	☐ 6

8. **COMMUNICATION EFFECTIVENESS:**
During the period has communication been effective:

Occasionally	Half the time	More often than not	Usually	All the time	Score
☐ 1	☐ 2	☐ 3	☐ 4	☐ 5	☐ 6

9. **COMMUNICATION FRANKNESS:**
During the period has communication been frank:

Occasionally	Half the time	More often than not	Usually	All the time	Score
☐ 1	☐ 2	☐ 3	☐ 4	☐ 5	☐ 6

10. **PUBLIC ISSUES:**
The extent to which negative feed back from the public has been successfully dealt with during the period (environmental, traffic safety, private property, noise):

Occasionally	Half the time	More often than not	Usually	All the time	Score
☐ 1	☐ 2	☐ 3	☐ 4	☐ 5	☐ 6

Total score for period = { }

Comments: ____

____

____

____

____

____

____

Signed: ______________________ **Date:** ______________________

Figure 34. (*Continued*)

ANNEXURE 8

DAIKYO

GREEN ISLAND RESORT PROJECT
CAIRNS • AUSTRALIA

PARTNERING ASSESSMENT SUMMARY

for period ending date 22/10/93 Period No 6

	CUMULATIVE SCORES					TOTAL SCORES	NO OF MEMBERS	AVERAGE SCORE
	1	2	3	4	5			
1. Work			9	56	10	75	19	3.95
2. Programme			18	36	20	74	19	3.89
3. Safe Job		2	3	48	25	78	19	4.11
4. Team			18	52		70	19	3.68
5. Future Work			18	44	10	72	19	3.79
6. Communications		2	24	40		66	19	3.47
7. Issues			24	48		72	19	3.79
8. Quality			12	44	20	76	19	4.0
9 Community/ Environment			6	40	35	81	19	4.26
10. Operations			12	44	20	76	19	4.0
11. Financial		6	33	12	10	61	19	3.21
								42.15

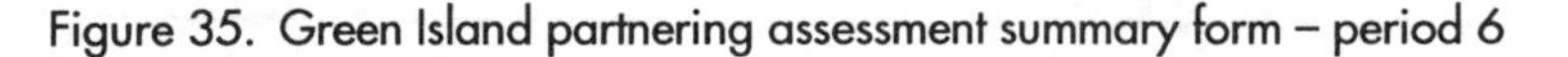

Figure 35. Green Island partnering assessment summary form – period 6

dealing with difficulties such as drawing delays or drawing change. It also saved the totally abortive post-contract dispute time and cost.

Author's comments

It may be that the logistic difficulties perceived by the principal(s) emphasized the need for a good strong project organization structure and relationships, and so prompted the formalization of the partnering process. In any event the principles of good strong project management for such a project, as well as for any other, are always needed for success.

Focus on the particular need for good relations on all one-off construction projects (are there any others?) brought about by use of the latest thinking, know-how and techniques of quality management will encourage the teamwork and so the greater likelihood of complete success and satisfaction for all parties involved.

ANNEXURE 9

THIESS

GREEN ISLAND RESORT PROJECT

PARTNERING ASSESSMENT SUMMARY CAIRNS • AUSTRALIA

for period ending date 22/10/93 Period No 6

No of forms for the period 19

	Average Scores	Average Score last period	Variance exceeds 3 scale points	
			Yes	No
1. Work	4.05	3.95		x
2. Programme	3.39	3.89		x
3. Safe Job	4.22	4.11		x
4. Team	3.72	3.68		x
5. Future Work	3.72	3.79		x
6. Communic-ations	3.5	3.47		x
7. Issues	3.22	3.79		x
8. Quality	3.94	4.0		x
9. Community/ Environment	4.39	4.26		x
10. Operations	4.22	4.0		x
11. Financial	3.44	3.21		x
Total	42.15			
Overall average	383.2			

Action Plan:	Action By:

TREND by PERIOD

500
450
400
350
300
250
200
150
100

Period 1st 2nd 3rd 4th 5th 6th 7th 8th 9th 10th 11th 12th

Figure 36. Green Island report and action plan

Case study no. 12 (civil engineering)

Sunshine Coast Motorway Stage 2 – Queensland, Australia

The project

This work was covered by four separate contracts to pre-qualified contractors:

Area 1.	Motorway	September 11, 1992 to Thiess Contractors
Area 2.	Motorway Centre	November 27, 1992 to John Holland
Area 3.	Motorway North	November 13, 1992 to Abigroup
Area 4.	NW Arterial	September 24, 1992 to Leighton Contractors

It was originally intended that contracts for Areas 1 and 4 would not be released until early 1993 because they were of a shorter duration but this decision was changed to allow earlier job opportunities on the Sunshine Coast. This has had the result that Areas 1 and 4 were complete several months earlier than the rest. It created a minor problem with Area 1 because the pavement condition needed to be maintained until Areas 2 and 3 were complete. It could not be put in service until they were complete.

The contract was expected to run 40 weeks at the outset. In or about May 1993, Area 1 contract achieved practical completion.

All Contractors made good use of a relatively dry summer and got most of their work up out of the ground before the wet season.

In Area 1, difficulties arose for Thiess Contractors (an active promoter of partnering in Australia) while trying to drive 35 metre long piles through a 2 metre layer of 'coffee' rock, silty clays and medium dense sands to the required founding levels. The piles were in segments. There was an 18 metre long 500 millimetre prestressed concrete (PSC) pile with two 350 millimetre square 'Balkan' piles with mechanical splices giving a total pile length of 35 metres. The pile geometry, the high water table and the varying resistance of *in situ* materials made it difficult to reach founding level without over-stressing these piles. The most significant variation was one for A$50,000 (£23,000) arising out of the difficulties with these segmental piles.

The partnering conference

The partnering, or 'kick-off', conference was held shortly after work had started. Because the work could not be left unsupervised the conference was held on-site and lasted about a day. A consultant 'facilitator' was employed. The outcome was a partnering charter which read:

> 'We, the Partners in the Sunshine Motorway Project – Stage II, are committed to achieve the performance and relationship objectives conjointly prepared and under-mentioned:

(*a*) performance goals/objectives:

- quality job – fit for purpose
- timely completion
- safe job
- minimize cost to all parties
- no unresolved disputes at completion of contract
- commitment to ADR
- timely exchange of information.

(*b*) relationship objectives:

- team understanding of work practice and restrictions of all members
- maximum satisfaction, i.e. no dissatisfied parties including individuals, principals, engineer, contractor and community
- effective and frank communication process
- early and fair resolution of issues
- harmonious workplace and close relationships between all parties and individuals on this job and after
- development of mutual respect for the competence of all parties
- commitment to protect the rights of the public'.

The persons invited to the initial conference were participants in the project down to and including foremen. Sub-contractors were not included because at that time no sub-contractors had been appointed.

There were about six sub-contractors, and Thiess were responsible for their performance. There was no later induction into the charter for sub-contractors once appointed. (Note, the separation of performance goals from those relating to human relationship objectives.)

The conference established:

- ownership of the charter
- increased communication
- effective communication
- harmonious relationships.

The client's project manager thought the concept of partnering had the potential to modify behaviour. While sceptical of its effect, he did everything to police it.

The issue resolution system

It was recognized that resolution of issues was a critical factor, and an 'issue resolution system' was established which required:

- problems to be resolved at the lowest level
- unresolved problems could be escalated upwards by both parties

and this should be done before project delays and costs were caused
- there should be no jumping levels of authority
- ignoring the problem or not making a decision was specifically stated to be unacceptable.

Joint evaluation

A joint evaluation process was also a main concern. It was agreed that monitoring should be simple and take little time so two forms were developed:

- an assessment form
- a report and action plan.

The seven performance goals and objectives and three of the relationship objectives were monitored. These were:

- early and fair resolution of issues
- communication effectiveness
- communication frankness.

A partnering assessment form was completed by each of the participants prior to each site meeting. The results of that assessment were then translated to a partnering report and action plan.

Final evaluation

This was the 'close up' conference. This final review concentrated on what the parties would like to be done better next time, either what each party could do better themselves or what each party would like the other party to do better next time.

The project manager's superintendent was not sure how far partnering had contributed to the success of the contract. His view was that a good contract should be put ahead of partnering and that partnering was best regarded as an enhancement to the management of the contract. He also thought it important that the parties had previously worked together.

Author's conclusion

The results of partnering this project seem modest but worthwhile. The project closed ahead of time and 5% under budget with no major outstanding claims or issues.

The procedures and their results appear to dwell on the more mechanistic aspects of partnering. The success, as compared with the imposed system on the Macquarie University Science Block, perhaps owes much to the initiative coming from the parties and their real ownership of the charter and process. The 'soft' or cultural aspects developed within a conventional civil engineering ambience and the logistical implications and complexities may be simpler than on many building projects.

9 Conclusions: the way ahead for stakeholders

Competition brings out the best in products and the worst in people.
David Sarnoff 1891–1971

Project partnering is a realistic and practical outcome to developing teamwork through a totally quality managed approach for individual building and civil engineering projects.

- It must start with a commitment from top management in all the companies involved from the conception through construction and commissioning to completion.
- This must include the client's top management with the responsibility for and the authority over the project. Without commitment at that level the conviction will not be developed and effective communication to all areas and to all levels involved in the project is unlikely to take place, and any commitment obtained initially will not be maintained.
- The partnering charter and the knowledge of all that is involved in the principles and practice of partnering is essential to its success.
- Education in these two facets are among the essential tasks for the initial workshop on every project, even where it is believed that the firms involved are aware of or are committed to the principles from previous operations, because every project will have different people involved. Even if some come from corporate cultures that are known, it is certain that for many years to come there will always be cultures which have yet to embrace the concept or commit to the quality view that there is a better way and that this is itself subject to continuous improvement.
- The intention to partner should be conveyed by the client from the very conception of the project. Once the feasibility study is positive – so should the partnering mechanisms be initiated to confirm its possibility and practicality.
- Assessment and evaluation by measurement against benchmarks should be made at pre-determined stages.
- A final review to provide feedback for all involved should always be carried out. These will set new benchmarks for improved performance

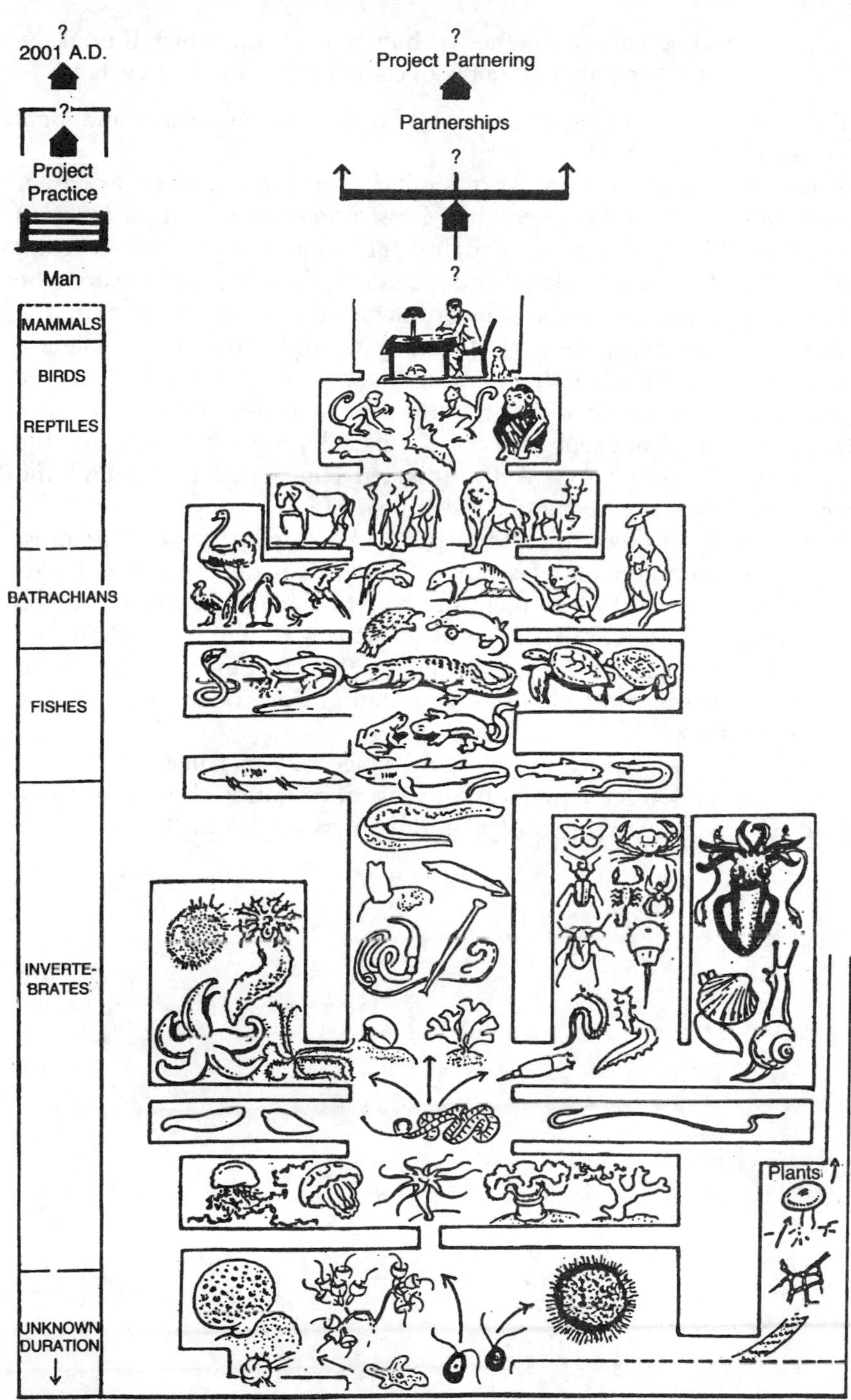

Figure 37. Continuous evolution and improvements?

on the next operation, whether with the same team, which is unlikely, or with a newly created assembly of companies to forın a new team.

But, how to achieve these conclusions in the face of a confrontational climate and consolidated cynicism?

These conclusions are not so dissimilar from those arrived at earlier (and contained in *Total Quality in Construction Projects*, pp. 153–156) which concluded that improved quality performance for the construction industry must be project related and include the whole project team – the manufacturer, the specialist sub-contractor, the main contractor, the professional designers, the project managers, and above all the client – the customer the whole industry exists to serve – all of whom must be involved in the process and through this process achieve Deming's 'joy of work' that comes from a job well done. This will put back the fun and the excitement that comes to a well motivated team creating the building from a clean sheet of paper and a green-field site.

The secret is through imparting greater knowledge to all. This must come from education and learning. The construction industry learns afresh on every project – habitually, but regrettably too frequently (and too little) from the mistakes that have been made – by the other chap. But construction remains and must always be the outcome of a combined operation, a team effort of inter-dependence, not one torn by eternal internecine strife.

Partnering, the sharing of aims and objectives in construction, is a culture that mankind after millions of years of evolution should surely be able to achieve even in the construction industry by AD 2000!

Appendix 1
Sample documents and model forms for partnering

Contents list for documents in Appendix 1

The effective development of a 'partnering' culture on construction projects will be assisted by including reference to the concept through the following 'model' documents produced by TQM Polycon.

Module D.212.B
11/90

QUALITY MANAGEMENT AUDITING
2nd Party Project Audit

CLIENT (Project Manager, Architect, Engineer, Surveyor) / GENERAL CONTRACTOR

1. The purpose of this Project Questionnaire is to establish the comprehensive effectiveness of your management methods, systems and people to 'GET IT RIGHT FIRST TIME' economically and effectively. It will also have regard to your company's contracting and legal procedures with particular emphasis on measures to prevent contractual disputes arising.

   This audit is concerned with the totality of the factors that affect your company's Quality Management and could therefore be said to be of its 'Quality Management System', ie: the organisation structure, responsibilities, activities & resources and events that, together, provide organised procedures and methods of implementation to ensure the capability of your firm to meet our specific project quality requirements.

   Our aim is to develop a 'Project Quality Plan' which is a documented set of activities, responsibilities and events serving to implement its quality system.

   The Quality Plan must of necessity embrace all those involved in the project from the time of commencement of the development. It will therefore include the work of the design team and the interface with Contractor and Sub-Contractor. It must also embrace the activities, operation and decisions made by or on behalf of 'the Client'.

   A similar audit was carried out at the inception phase of the project as the basis for the initial team building exercise and, it is anticipated, there will also be audits conducted on principle Sub-Contractors and suppliers.

2. This 2nd Party Audit, whether carried out by a Polycon consultant or by a project manager, is aimed at:-

a) identifying the objectives and scope of the systems to be established;

b) determining the present stage of development of your quality procedures and their documentation;

c) providing a report for discussion by the quality policy or steering group, which will also provide the basis for the project quality programme leading up to final assessment / selection.

3. This questionnaire is intended to help in the selection of either professional organisations, or a Contractor/Sub-Contractor, through whom the Client wishes to carry out development or building works. This audit will always be limited to those aspects of the audited company relevant to the Client's specific project needs.

4. Many matters involved in the project quality system will affect more than one department or area of management responsibility and this questionnaire is intended to provide information that will aid selection. It is likely that the person receiving this may not be familiar with all aspects of his company, and so the questionnaire will form a good checklist for him also.

   This questionnaire is therefore divided into sections which may have some elements of overlap where activities are primarily the responsibility of one function and of the management related to this function.

   A. General Organisation - which should normally be completed by the Managing Director, Senior Partner, General Manager or Company Secretary.

The following sections, where appropriate in a larger organisation, should be completed by the appropriate Departmental Manager.

B. Marketing
C. Cost Control
D. Contract Management
E. Buying
F. Defects and Potential Claims
G. Client Company General Support and Administration
H. Subsidiary Companies

PRELIMINARY INFORMATION **2PQ NO. ....**

(to be retained by Audited Company)

Client Name: Address:	Tel:	Fax:
Project Title & Reference		
Client's Project Manager/ Representative		
Nature of Project **(attach any documents that you think will assist)**		
Intended Timing of Project's Main Phases		
Likely Order of Cost: **Total Building Cost** **Any Phases:**	£ £	
Likely order of cost of the particular package of work that this Audit covers	£	
Preferred timing of this work package		
Date for return of this Questionniare	/ /	
Date for placing Letter of Intent (Contract/Order)	/ /	
Dates for Commencement of work (covered by this package)	/ /	

QUALITY MANAGEMENT AUDIT

A. GENERAL ORGANISATION

1.	Name of Organisation and Address Telephone No. Fax No.	
2.	Client Contact - Name - Tel.	
3.	Is this part of a larger group or are there subsidiary companies/locations? If yes, name and address of group and/or subsidiary(s).	
4.	What is the approximate number of employees in the Company(s) directly involved in the project which is the subject of this Audit	0 - 49 50 - 99 More than 100
5.	Please attach a list of current projects, their location and the approximate duration and contract value of each.	
6.	Do you have an organisation chart? Please attach a copy, together with Management Responsibility Mandate or Job Description pertaining to this Project	YES/NO

QUALITY MANAGEMENT AUDIT

A. GENERAL ORGANISATION (2)

7.	**When was your firm founded?**	
8.	**Is the founder still in the business?**	
9.	**How many Partners/Directors are there?**	
10.	**Does the Company have a legal adviser on the staff? Or, do you use a regular Solicitor? If so, please give name and title.**	
11.	**How often do the principals meet to conduct the overall business of the firm? Is this a regular event?**	
12.	**Please attach an organisation chart for this project, together with the CV's of the project manager/director and his site manager**	

QUALITY MANAGEMENT AUDIT

B. MARKETING

B.1 **Please list recent projects of a similar size and nature to the project which is the subject of this Audit and which can be visited for an assessment of quality, together with the names of the client(s) to whom reference can be made in each case.**

COMPLETED BY: ....

NAME **JOB TITLE**

QUALITY MANAGEMENT AUDIT

C. COST CONTROL

C.1 **Please describe your procedures for preparations and reporting of individual contracts, profit centre costs and variances to budgets and forecast - attach sample report. Please indicate if there will be any varience with these procedures for the project which is the subject of this Audit.**

COMPLETED BY: ....

NAME **JOB TITLE**

QUALITY MANAGEMENT AUDIT

C. COST CONTROL (2)

C.2 **Approximately what percentage of turnover is carried out under the following types of construction and forms of contract/sub-contract. Insert value if this is easier than expressing as a percentage.**

	JCT 80	JCT 81 with Design	IFC 84	JCT 87 Mgmt Cont.	JCT Minor Works	Sub-contract Form	Construction Mgmt.	Mgmt. Contr-acting	Other
A. Private Sector - New Build									
- Repair, Maint. & Improvement									
- Housing									
B. Public Sector - New Build									
- Repair, Maint. & Improvement									
C. Own Development - New Build									
- Repair, Maint. & Improvement									
D. Other									

NOTES:

COMPLETED BY: ....
NAME **JOB TITLE**

QUALITY MANAGEMENT AUDIT

D. CONTRACT MANAGEMENT

D.1 **Briefly describe your procedures, including a short method statement, for running the project which is the subject of this Audit. Your method statement should also include a programme indicating the time-span of the works and highlighting any critical interface with Sub-Contractor, Supplier, etc.**

D.2 **Do you normally prepare 'Release of Information' Schedules?** **YES/NO**

If Yes, please submit a recent sample.

D.3 **Do you normally prepare 'Restraints Programmes' for your Sub-Contractors/Suppliers?** **YES/NO**

If yes, please submit a recent sample.

COMPLETED BY: ....

NAME **JOB TITLE**

QUALITY MANAGEMENT AUDIT

E. BUYING

E.1 Please describe briefly below your procedures for procurement, management and payment of sub-contractors and suppliers -and attach examples of any forms used with particular reference to this project.

E.1 Do you normally "Audit" your Sub-contractors/Suppliers? YES/NO

If yes, please state the procedures normally adopted and who will undertake such an Audit.

COMPLETED BY: ....
NAME JOB TITLE

QUALITY MANAGEMENT AUDIT

F. DEFECTS & POTENTIAL CLAIMS

F.1 **Please outline your policy and any procedures for handling claims and dispute situations leading up to legal action, claim and arbitration (for and against)**

F.2 **How many disputes have you had in the last 3 years that have gone to:**

Adjudication	**No.**	**Amount £**
Arbitration	**No.**	**Amount £**
Litigation	**No.**	**Amount £**

F.3 **Do you have any existing or outstanding disputes?**

Approximately how many, and give approximate value.

YES/NO	**No.**	**Value £**

COMPLETED BY: ....
NAME **JOB TITLE**

QUALITY MANAGEMENT AUDIT

G. GENERAL SUPPORT & ADMIN.

1. Internal Communications **Do you have a news letter/company newspaper or similar? What is it called?** **How often is it distributed?**	
2. Who is your Safety Officer?	
3. How do you handle security matters on site?	
4. Who keeps records of claims made and notified to insurers?	
5. Have you ever been accused of unfair dismissal by any employee? or of Sexual or Racial Discrimination? **6. What was the outcome?**	**YES/NO WHEN?**

COMPLETED BY: ....

NAME **JOB TITLE**

QUALITY MANAGEMENT AUDIT

H. SUBSIDIARY COMPANIES AND
OTHER TECHNICAL SUPPORT SERVICES

H.1 Please list and describe the role of any subsidiary or associated companies or activity or technical discipline which are an important part of your company's activities, which are not covered elsewhere in this Questionnaire. If you can please give some indication of the scale of activity, e.g. function, number of people employed, etc.

Title:

Scale of Operations

Location

etc.

COMPLETED BY: ....
NAME JOB TITLE

PROJECT QUALITY AUDIT

1. **Having completed this Questionnaire, which addresses many factors and areas of your firm's organisation in order to assess your suitability for this project, will you please indicate below dates when it will be possible for the Auditor(s) to visit your firm to discuss this Questionnaire, and any other factors he may deem necessary, with the appropriate members of your staff, who should include key staff responsible for the project.**

2. **The Auditor may visit our offices:-**

on ____________________ **between** ____________ **and** ____________

OR

on ____________________ **between** ____________ **and** ____________

OR

on ____________________ **between** ____________ **and** ____________

3. **Have you any other comments which may have a bearing on this Audit?**

Signature: ________________________ **Date:** ______________________

Position: ____

Please return the completed Questionnaire to the Auditor using the "Private & Confidential" label attached.

HAVE YOU TAKEN A COPY?

Project specification – sample provision

Partnering. The Building Owner intends to encourage the foundation of a cohesive partnership with the Design Team, Contractor and its sub-contractors. This partnership will be structured to draw on the strengths of each organisation to identify and achieve reciprocal goals.

The objectives of this approach are effective and efficient contract performance, intended to achieve completion within budget, on schedule, and in accordance with plans and specifications.

Participation in this partnership will be totally voluntary. Any cost associated with developing this partnership will be agreed to by both parties and will be shared equally with no change in contract price. To implement this partnership initiative, it is anticipated that within x days of Notice to Proceed the Contractor's on-site project manager and the Owner's on-site representative will attend a partnership development meeting followed by a team-building workshop to be attended by the Contractor's key on-site staff and Owner's personnel and design team. Follow-up workshops will be held periodically throughout the duration of the contract as agreed to by the Contractor and Owner.

An integral aspect of partnering is the resolution of disputes in a timely, professional and non-adversarial manner. Contract Management Adjudication, an alternative dispute resolution methodology will be encouraged in place of the more formal dispute resolution procedures. Contract Management Adjudication will assist in promoting and maintaining an amicable working relationship to preserve the partnership. In this context CMA is intended to be a voluntary, non-binding procedure available for use by the parties to this contract to resolve any dispute that may arise during performance.

B

Model pre-bid partnering letter to managing directors on the bid list

John Abel, Esq.
Quality Build Construction Ltd.
New Town
Blankshire.

SUBJECT: Partnering on XYZ Project

Dear Mr. Abel,

I understand your company intends to bid on this contract. I am making a special effort to inform senior executives of all interested companies of an exciting new concept in the management of this project. It is my intention to establish a formal "Partnering" agreement and programme with the successful bidder.

Partnering is a process promoting teamwork, minimising confrontation and, hopefully, eliminating the need for litigation, where all stakeholders finish the job a winner. It is a challenging endeavour that requires the commitment of senior management.

I will present the details of Partnering during a pre-bid conference luncheon on (date) here in (location). I hope you will have your representative report back to you with the particulars. You are certainly welcome to have one of your executives attend the luncheon to participate first hand.

Please feel free to contact me if you have questions at (telephone number). Reservations for the luncheon should be phoned in to (appropriate person) by (date).

Yours sincerely,

Owner's Chairman or Managing Director

C

Sample letter to awardee requesting a meeting to discuss the partnering concept

John Abel, Esq.
Quality Build Construction Ltd.
New Town
Blankshire.

Dear Mr. Abel,

Congratulations! I was delighted to find (company name) made the apparent lowest bid on the project. Your Company has a reputation for excellence and we look forward to a mutually rewarding relationship.

I hope to have all the administrative formalities completed by (date) when I will make the formal contract award. In the meantime, I would like to propose a meeting with you shortly, at your headquarters, to discuss a "Partnering" approach to managing the project.

My concept of partnering is recognising shared risk and common objectives, promoting cooperation, minimising confrontation and eliminating litigation. Success will be when all stakeholders finish the job a winner. It is a challenging endeavour that required up-front agreement on expectations, helpful systems and, most importantly, the unqualified commitment of senior leadership.

I look forward to hearing from you to determine when a meeting may be convenient.

Again, congratulations and best wishes!

Yours sincerely,

Owner's Chairman/Managing Director

D

This is an agenda for a simple one-day Partnering workshop. For larger projects the parties may wish to expand the time and scope of the workshop by including discussions of problem solving styles, prior experiences, risk management philosophies, anticipated difficulties, and/or simply give more time for the parties to become better acquainted - in small or larger groups.

Workshop

Agenda

9.00–9.15 a.m.	Opening Remarks of Senior Executives – Why we are here.
	Client – Contractor – Others
9.15–9.30 a.m.	Introductions
9.30–10.30 a.m.	Partnering Overview (by Facilitator/ Project Manager/Employer's Representative?)
10.30–10.45 a.m.	BREAK
10.45–11.15 a.m.	Exercise No. 1
	Barriers, Problems, Opportunities
11.15–11.45 a.m.	Report and Discussion in Entire Group
11.45–12.00 a.m.	Develop Project Mission Statement
12.00–1.00 p.m.	LUNCH
1.00–1.15 p.m.	Exercise No. 2
	Interest, Goals, Objectives
1.45–2.15 p.m.	Report, Discussion, Identification of Common Goals and Objectives
2.15–2.30 p.m.	BREAK
2.30–3.15 p.m.	Exercise No. 3
	Issue Resolution and Team Evaluation
3.15–4.00 p.m.	Report Discussion, Agree on Process and Format
4.00 p.m	Sign Charter

E

Initial partnering meeting exercises

For these exercises the parties will break into two groups (Owner and Contractor). These questions are answered and then reported back to entire group. Discussion facilitates understanding. Design Team members should divide between the two groups.

Exercise No. 1 – barriers, problems and opportunities

- What actions do the other groups engage in that create problems for us?
- What actions do we engage in that we think may create problems for them?
- What recommendations can we make to improve the situation?

Exercise No. 2 – interest/goals/objectives

- What direct and indirect interests do we have in the outcome of this project?
- Given our interest, what are reasonable, achievable goals for which we can strive?
- What specific measurable objectives can we identify that move us toward our goals?

(Again, the parties separate into Owner and Contractor groups. When results are reported back to the entire group, common objectives emerge. From these, a specific list of charter objectives are developed along with the group mission statement.)

Exercise No. 3 – issue resolution/team evaluation

- What should our issue resolution policy require?
- How should the issue resolution process work?
- What are the roles and responsibilities for all levels of the partnership in issue resolution?
- What should be the invocation procedure for Adjudication?
- How can we evaluate the progress of the partnership in achieving our goals and objectives?
- Who initiates the evaluation, who has input to the evaluation, and who sees the evaluation?
- What actions should the evaluation trigger?
- Should the evaluation process include follow-up workshop(s)?
- If so, when and who is responsible?
- Who should attend?

(This exercise may be conducted in one large group discussion. Specific follow-up tasks may be assigned to ensure closure on procedures and evaluation forms.)

F

Partnering charter

The mutual goals and objectives of the stakeholders form the Partnering charter. The charter for each project, therefore, will be unique to that project. The charter may be a simple statement about communication and cooperation in all matters affecting the project and resolution of conflicts at the lowest level. However, the following provides an idea of objectives which might be included in a charter.

Partnering charter for (project)

We are a team dedicated to providing a quality project in accordance with the contract. We are committed to both employee and public safety, protection of the environment, and minimising inconvenience to the public.

1. Communication objectives

We intend to deal with each other in a fair, reasonable, trusting and professional manner. We will:

(i) Communicate and resolve problems within the terms of the contract.
(ii) Make decisions at lowest possible level.
(iii) Conduct open, honest communication.
(iv) Treat each other with mutual respect, resolve personal conflicts immediately, and avoid personal attacks.
(v) Give timely notification of meetings.
(vi) Punctually support the weekly and progress meetings.
(vii) Not allow grudges to interfere with professionalism.

2. Conflict resolution system

Step 1: It is preferred that conflict be discussed and resolved at the level at which it originates.
Step 2: When conflict is not resolved at the originating level, it will be quickly taken to the appropriate level of management,or
Step 3: Submitted in accordance with the procedure for adjudication

3. Performance objectives

(i) Complete the project without litigation.
(ii) Utilise cost reduction incentive proposals.
(iii) No delays to project.
(iv) Project completed on time.
(v) No lost time due to injuries.

(vi) Promote positive public relations.
(vii) Provide safe passage of the public through the project.
(viii) Make project enjoyable to work on.
(ix) Construct and administer the contract so that all parties are treated fairly.

We, the undersigned, agree to make a good faith effort to undertake and implement the above as applicable to each of us:

General contractor personnel	Owner Personnel
______________________	______________________
______________________	______________________
______________________	______________________
______________________	______________________
	Architect/Engineers
______________________	______________________
______________________	______________________
______________________	______________________
______________________	______________________
______________________	______________________
	Key Suppliers
______________________	______________________
______________________	______________________
______________________	______________________
______________________	______________________
______________________	______________________
______________________	______________________
______________________	______________________
______________________	______________________
______________________	______________________
______________________	______________________
______________________	______________________
______________________	______________________
______________________	______________________
______________________	______________________
______________________	______________________
______________________	______________________

Dated for Partners / /

H

Contract management adjudication

Despite the successful working within a Partnering Charter sometimes unforeseen events occur - may be in the condition of the groundworks or unilateral changes in the organisation of one or more of the participating companies.

Particularly where technical issues are involved and where the understanding of the implications is foreign to one of the parties it may be beneficial and quicker to refer the issues to top management who may (but only may) be sufficiently outside the operations to see the issues clearly and be able to resolve them within the concept of the process. It may also be better for them to make their own assessment and then to refer the unresolved issues for a third party adjudication and award, binding upon the parties for the remainder of the project's contract, but not final until the end of that contract when all parties could then participate in a final evaluation and, if necessary, in the finalisation of the Award by the Adjudicator - all within the spirit and concept of "Partnering". Top management for the parties could then present, or advocate their views on the issues to assist the adjudicator to come to the right decision (i.e. in their favour!)

For this reason (and it could very much simplify the actual drafting of the Contract document) it would be sensible to have a "Contract Management Adjudication" clause drafted into the Partnering Charter (and/or in the Specification and the Contract Documents in the following terms:

> "In addition to/substitution for the provision of Arbitration under Article 4 of the Recitals, Contract Management Adjudication procedures shall apply to disputes or differences arising out of or in the course of the Works.
>
> It is also intended that of shall be appointed jointly by the Employer and Contractor to resolve any dispute or difference that may arise during the progress of the Works between the Employer or his Architect, Engineers, Quantity Surveyor or Supervising Officer and the Contractor or any Sub-contractors, to issue Directions for the continuance of the Works under these conditions but without prejudice to the right of any party to seek formal Arbitration if they do not accept the Adjudicator's Award as final and binding upon them all at the end of the Contract.
>
> The Contractor will be required to incorporate the provision for Adjudication in any sub-contract for the Works and for the supply of items for the Works whether these are from nominated persons or otherwise."

It shall be a condition of any reference to the Adjudicator that the parties shall have first used the joint formal evaluation and issue resolution steps in the Partnering Charter and shall submit these in writing to the Adjudicator who will then decide and conduct the next steps in the resolution process up to the making of his written Adjudicator's Award.

J

Partnering programme - project status evaluation (Project name)

This form to be filled out by job-site partners prior to monthly progress meeting. Completed form is distributed at progress meeting and its review becomes the last agenda item of this meeting.

Date:

Item	Evaluation		
	Contractor	Owner	Others
1. Quality of Project			
2. Resolution of Jobsite Problems			
3. Tone of Communication			
- Progress Meetings			
- Letters			
- Oral			
4. Special Reports Required			

Name of Partnering Representative: ....

K

Model evaluation form for partner rating

Project: ______________________________

Partner evaluated: ______________________________

By: ______________________ Date: ______________________

Consideration of partnering factors:-

		FACTOR	SCORE
1.	ADVERSARIAL/AGGRESSIVE/AVOIDANCE/SELF-INTEREST	[ ]	( )
2.	ACCOMMODATING/COMPROMISING/SOME POSTURING	[ ]	( )
3.	SYNERGY THROUGH COLLABORATIVE/WIN-WIN/TEAM INTEREST	[ ]	( )
4.	OVERALL TRUST/CANDOUR	[ ]	( )
5.	ARE THE PARTNER'S COMMUNICATIONS:-		
	° Open, Honest, Timely?	[ ]	( )
	° Active and cooperative?	[ ]	( )
	° Is the number and tone of letters good?	[ ]	( )
6.	PROBLEM SOLVING ACTIONS:-		
	° Win-Win Synergy	[ ]	( )
	° Solved at Lowest Level?	[ ]	( )
	° Speed of escalation when beyond authority?	[ ]	( )
7.	PROGRESS ON GOALS	[ ]	( )
	Overall Score	[ ]	( )

Project-specific issues and comments

1. ______________________________

2. ______________________________

3. ______________________________

4. ______________________________

Insert Factor [A] = Excellent and Score (4)
[B] = Good and Score (3)
[C] = Adequate and Score (1)
[D] = Poor and Score (0)

L

Alternative model evaluation partner rating form

PROJECT NAME : ....

PARTNER EVALUATED: ....

EVALUATED BY: ....

	LOW 0	GOOD 1	EXCELLENT 3	COMMENTS
QUALITY	0			
BUDGET PROGRESS	0			
SAFETY	0			
COMMUNICATION	0			
TRUST/CONFIDENCE	0			
TEAMWORK	0			
ISSUE RESOLUTION	0			
ADMINISTRATION	0			
VALUE ENGINEERING	0			
COMMUNITY RELATIONS	0			
✓as appropriate and add	0			Total

Observations and Suggestions for improvement:

Date: / /

Signature: ....

M

Partnering report and action plan

for week/fortnight/month* ending date __/__/__ Period No. ______ (* delete)

Average score from Partnering Assessment Form No. of forms for the period ______

(#please tick)

	Average Score	Variance exceeds 3 scale points # Yes	No
1. Quality			
2. Program			
3. Safety			
4. Cost			
5. Information			
6. Workplace			
7. Resolution of Issues			
8. Communication Effectiveness			
9. Communication Frankness			
10. Public Issues			

Total ______

Overall average ______
(Total/10x100)

Action Plan:	Action By:

TREND by PERIOD

500												
450												
400												
350												
300												
250												
200												
150												
100												
Period	1st	2nd	3rd	4th	5th	6th	7th	8th	9th	10th	11th	12th

Appendix 2
Survey on project partnering in the UK carried out in January 1995

The purpose of the survey was to establish the degree of interest in partnering and its practice in the UK, to encourage its use and to feedback the information gained to all those whose progress in quality management will allow them to benefit from its use on their own projects.

One-hundred-and-fifty organizations were approached in the initial survey, 50 of the largest client firms, the 50 largest general contractors, and another 10 or so who were thought to be active in this area, the balance was made up from the largest architectural, civil engineering and surveying practices.

The numerical results are given against the questions posed in the following pages. The response was poor, probably a reflection of both the stretched state of the industry pressurizing top management, and the low priority accorded to the subject.

Since project partnering in construction must originate at the top of the organizations involved the questionnaire was addressed in each case to the chief executive, senior partner, etc. Where the chief executive him- or herself dealt with it the response was quick, complete and informative. The total number of responses was 23 (15.3%). The figures given are as a percentage of those replies.

Fifty-two percent of the replies came from the named addressee; 26% from other appropriate top management, and 22% from quality managers, professionals, etc.

Only nine companies claimed to have participated in a partnering project, all of them initiated by the client but in only four of these were the key technical elements for success – a partnering workshop and a partnering charter – carried out. In two of these complete success was claimed, three achieved 66%, five felt some success was achieved, whilst one felt it was too early to judge.

Even within this small group there was a disparity of views on the suitability of projects for partnering. Most thought projects of from £500,000 and upwards were suitable but, strangely, there were many who thought that projects that were:

- complex
- fast track

- one-off (what others are there?)
- with limited definition of requirements

were unsuited to the approach. In the author's opinion these are the very projects that are most suited to the philosophy, ethos and technique of partnering.

This initial summary suggests that the UK is well behind both the USA and Australia in the practice of partnering and further, as was found in Australia at the beginning of 1994, there is some interest but very limited knowledge of both the principles and the practical application of the techniques.

The one question that invited the survey to reject the whole concept as unworkable, inadequate or impractical in the UK, at least at the present time, did not draw as much fire as was expected, probably because the survey was sent only to the largest (and best) organizations. Nevertheless 43% did respond and there was a well-spread range of views by those who would not accept the ethos.

- 30% thought the client would not
- 4% the contractor
- 7% the financiers
- 13% the sub-contractors
- 13% the quantity surveyor.

One respondent generally well clued up on the approach summed it up with 'the barrier is attitude not types of organization'.

A. EXPERIENCE of Project Partnering

A. Please indicate your role in the Industry:

Confidentiality of all individual responses will scrupulously maintained

CLIENT	6	PROJ. MANAGER	1
ARCHITECT	1	CIV/STRUC. ENG	1
MECH.SERV.ENG.	1	QUANTITY SURVEYOR	1
GEN.CONTRACTOR	10	SPEC/SUB. CONT.	2

---

1. **Have you (are you) participated in a Partnering Project?**
   If YES, go to Question 3. If NO, go to Question 2.

YES	9	NO	10

2. **Have you considered doing so?**
   If NO, go to Question 11.

YES	5	NO	3

3. **Who instigated the concept?**

YOU	4	CLIENT	10
DES.TEAM	0	GEN.CONT	4
OTHER	1	Who	

4. **Was there a Partnering Charter?**
   With how many Objectives?

YES	5	NO	7
2 - 5	3	6 - 10	2
10 - 15	0	More than 15	0

5. **Was there a Partnering Workshop?**
   If NO go to Question 9

YES	4	NO	6

6. **When did it take place?**

After Site Commencement	0	Pre-Start on site/ Post Contract	4
Pre-Con. Signing	0	Other	1

7. **Who was involved?**

CLIENT	5	DESIGN TEAM	5
GEN.CONTRACTOR	6	SUB-CONTRACTORS	3

8. **At what level?**

CHIEF EXECUTIVE	2	TOP MANAGEMENT	5
OFFSITE MANGMT.	1	SITE MANAGEMENT	2

Down to what level? ....

9. **How many Stakeholders were involved in your project?**

Less than 4	7	4 - 6	1
7 - 12	1	Over 12	0

10. **Were/are the Objectives being achieved?**

Completely	2	More than 66%	3
Between 66 & 25%	4	Less than 25%	1
		None	

B. <u>OPINION ON THE CONCEPT & PRACTICE</u> of Partnering in Construction Projects

11. In your opinion who should initiate the the Project Partnering Concept?

General Contractor 26% Design Team 13%
Project Manager 9% Client? 78%

12. Who should be involved in the Partnering?

Client? 78% Lawyers? 13%
Project Manager? 52% Design Team? 56%
General Cont? 73% Some Specialist Conts? 61%
All Sub-conts? 17% All Site Workers? 13%

13. Who, or what, are the Stakeholders in the Project?

Several respondents did not understand the question. One included "Banks" most considered as Q.12

14.a What size Project is suitable for Partnering?

£25 - £100,000 9% £100 - £500,000 9%
£500,000-£2.5m 47% £2.5m - £10 m 65%
Over £10 m 47% ***All*** ***9%***

b **What kind of Project is NOT suitable for Partnering?** e.g. Size, Duration, Type of Project?

"Small", "1 off", "Very Short", "Complex", "None"

15. Does "Partnering" require different Form of Contract?

Different/New 21% Amended Standard 39%

or can it be used with :

JCT 30.4% D & B 21%
ICE 13% GC Works 9.2%
Construc. Mangmt. 13% Mangmt. Contracting 17%

16. When should "Partnering" be discussed?
Tick any number of boxes

After signing the Contract/Sub-contract? 0%
Before Signing Contract/Sub Contract? 4.3%
At time of Tendering? 13%
At time of Invitation to Tender? 35%
At commencement of Design Stage? 61%

Some other time? When? ***Feasibility*** ***39%***

17. By whom should Project Partnering be discussed?

Initially? ***Client and Main Contractor***

At the Commencement of Operations on Site? ***Client, Main and Sub-Contractors***

At any other time? If so, when?

18. Who should "own" the Partnering Charter?

Client & General Contractor - 1; Client & Sub-Contractor - 1; Client - 1; Everyone - 12

19. Who stands to benefit financially from Partnering?

Client 69% Design Team 52%
Gen. Contractor 65% Spec. Contractor 52%

20. Partnering will not work in the UK on building projects because it will not be accepted by:-

Client* / Contractor* / Architect* / Engineers* /

Lawyers* / Financier* /Sub-contractors* / Others*

C. KNOWLEDGE (of facets of Partnering)

21. **Which of these statements is true, false of partly true?**

Partnering:-	TRUE	PARTLY TRUE	FALSE
a) is a voluntary process	61%	26%	9%
b) is a Project Management structure	9%	43%	26%
c) requires imposition by the client	35%	21%	21%
d) is a project management process, not a legal system	44%	26%	4.3%
e) is a covenant to fair dealing and good faith	69%	13%	0%
f) requires outside consultant involvement - to set up	13%	26%	26%
- to facilitate & run Workshop	13%	26%	26%
- to monitor results	4.3%	30.4%	26%
g) is a standard technique for projects	4.3%	13%	61%
h) requires tailor-making to each project	39%	26%	13%

22. **Can Partnering be expected to deal effectively with?**

	TRUE	PARTLY TRUE	FALSE
a) cost over-runs	61%	26%	0%
b) time over-runs	61%	21%	0%
c) client changes to requirements	56%	26%	0%
d) unforeseen site problems	39%	39%	4.3%
e) resolution of disputes	43%	35%	9.2%

---

23. Response by:- Chief Executive 52%; Other 26%; Quality Manager 22%

24. Would you like to be informed of the outcome of this Survey? YES 83% NO 9.2%

25. Would you like more information/advice on Partnering? YES 61% NO 13%

26. Would you be interested in an Appreciation Workshop? YES 56% NO 21%

- Total Response: Clients 26%; Professionals 9%; Contractors 43.4%; Pilot Group 22%

Appendix 3
Partnering awards for excellence

The 1995 Marvin M. Black Excellence in Partnering Awards

Deadline: January 17, 1995
Project name:
Location:
Nomination submitted by:

(name and title)

(company – must be AGC member)

(address)

(phone and fax number)

I Project description
- public or private
- procurement method ☐ competitive bid ☐ negotiated bid
- type of work

(i.e. school, highway, waste treatment plant, highrise, etc.)
- size

(approximate dollar value)

- location

(i.e. urban, suburban, rural, etc.)

- who initiated the partnering process?

II List stakeholders

(use separate page if necessary)

III Objective criteria (explain each item in one or two paragraphs)
- was a partnering charter prepared? (include a copy)
- were the goals of the charter realized?
- what was the safety record? (lost workday cases)

- did the project come in, at, or under budget?
- did the project come in, on, or ahead of schedule?
- did you prepare an issue resolution procedure?
- how many claims were filed on the project (give number)

IV Subjective criteria (answer in one or two paragraphs)

- describe the partnering evaluation process on this project
- describe any unique motivational activities employed
- list any team-building activities
- identify any innovative ideas that evolved through the partnering process (cost savings, improved productivity, quality, etc.)
- discuss how the overall project quality was attained
- describe project relationships and on-going relationships with the stakeholders (supply testimonial letters if possible)

V Bonus points (answer in one or two paragraphs)

- explain any special adaptations of partnering to fit this particular project
- offer your ideas for improving the partnering process

VI Other items/other pertinent information (not to exceed five pages).

How winners are selected

Marvin M. Black Excellence in Partnering Awards are judged on the basis of both objective and subjective criteria. Parts III and IV should be answered in one or two paragraphs. Include one additional page summarizing why the project should receive a Marvin M. Black Excellence in Partnering Award. Please include colour photographs of the project and people.

Assemble all materials in a three-ring binder so the pages may be easily removed for photocopying. Tabs or sheet protectors are not necessary. other items that may be included with the entry include press clippings, letters of commendation and testimonials, programmes from special events related to the project or other related material. Other items should not exceed five pages.

The judges look for complete information as outlined – not for sophisticated or 'slick' presentation. Your own staff will be most familiar with the project and usually submit the best entries. This form may be retyped or reproduced in your own entry. It is important that your entry follow the same order as the form.

The Master Builders Association of Australia Partnering Award – a strategy for excellence (1993 and 1994 Criteria)

Submission

The submission requires information on:

(*a*) **description of project**
(*b*) **list of project team members**
(*c*) **list of the mutual objectives** – drawn from the partnering charter (also attach a copy of the charter) and provide supporting statements about the achievement of the project team against the measures that were set for each objective
(*d*) **quantitative data on results for the following:**

- have safety goals been achieved?
- did (or is) the project on/under schedule?
- did (or is) the project on/over budget?
- were value engineering savings made?
- approximately how many potential claims were avoided?

(*e*) **statements explaining partnering's contribution to the success or otherwise in the following:**

- improvement in the quality of the project
- improvement in the application of innovative solutions to the management of the project
- long-term goodwill with other team members
- morale of all project team
- communication between project members
- application of a structured approach to problem solving and decision making
- sensitivity to the collective needs of project members
- is the client satisfied with the progress of the project?

(*f*) **Provide a statement on the overall effect that partnering had in terms of changing the cultures of the organizations participating in the process.**

Finally we would appreciate your advice on how the partnering concept can be improved, and how it can be better promoted.

Eligibility and judging criteria

(*a*) In the spirit of partnering the nomination must be made with the support of all parties on the project.
(*b*) The award will be made based on information provided by the team concerning the progress that has been achieved against the objectives

that were set down in the project mission charter as well as the adherence to the basic principles of partnering which include:

- the development of commitment to partnering at all levels in the organization and on the project
- ensuring all stakeholder interests are considered and there is a commitment to a win/win strategy
- stakeholders working together to achieve mutual goals and objectives
- the development of helpful systems and procedures for implementing their mutual goals and solving problems
- conducting joint and continuous evaluation based on the objectives that have been set
- communicating in a timely fashion and creating a decision-making process that empowers people at the lowest possible level.

(*c*) The project team is asked to supply quantitative as well as narrative statements from stakeholders concerning performance of the team against the objectives.

(*d*) The judges would also wish to consider how partnering has helped to change the organization's corporate culture.

(*e*) The award is open to all sectors of Australian industry and nominations will be accepted from all projects regardless of whether the project is complete or still under way.

Appendix 4
Principles of total quality management

General Introductory Comment

The principles were developed during 1993 by a working party of the "Consultancy and Training Providers Group" of the British Quality Foundation.

They were then circulated to a wide cross section of UK companies and organisations. This evoked many replies, suggestions and comments. The principles have therefore been compared to the working practice of organisations both large and small, within manufacturing, processing and distribution sectors of industry and in several areas of public service. Leading authorities on Total Quality Management also contributed their encouragement and suggestions.

The draft principles were presented to a full meeting of Consultants and Trainers in October 1993, who further contributed minor amendments. The principles have been amended after careful consideration of all comments and suggestions, some of which were contradictory. This version is neutral, in that it is not biased towards any particular industry or strong point of view, other than that TQM will be an essential ingredient of this Nation's future.

The principles of Total Quality Management are neither definitive nor prescriptive. The title is now used to represent a vast number of different schemes and initiatives. There is no intention to either justify some of these, or to place others outside the area of Total Quality Management.

The principles set out represent the spectrum of cultural activity that will be the firm foundation of any successful organisation developing Total Quality Management. Individually they represent strands of concern that inform the Total Quality approach, whilst together they embody the vital principles that represent the best practice within organisations practising Total Quality Management.

To represent the wide variety of organisations adopting these principles and to express the spread of interests involved the principles are stated with alternative and complementary concepts grouped together.

The principles have been developed from first principles, to stand on their own merits. They can be seen to complement the UK/European Quality Award Model, providing description and insight behind the 9 award blocks.

PRINCIPLES OF TOTAL QUALITY MANAGEMENT

1 Highest priority

Total Quality must overtly be the highest priority of the *organisation / company / individual.*

Explanation and description of principle.

Whilst an organisation's exact definition and understanding of Total Quality will be unique, the consistent application of this concept must always take precedence over all other considerations.

2 Quality definition

Any definition of Quality must include *meeting / satisfying / conforming* to *agreed / negotiated* customer *needs / requirements / wants / expectations.*

Explanation and description of principle.

There are several active definitions of Quality in general use. The one used by an organisation must be understood by everyone within the supplier/customer chain of which the organisation forms part.

Many organisations attempt to exceed customer requirements and expectations.

3 Customer definition

The concept of customers includes *investors / employees / stake-holders / suppliers / the community* and every interpersonal relationship.

Explanation and description of principle.

An organisation's customers are wide ranging. Some will be inside the organisation, others will be external. Some may be corporate, whilst others will be individual people.

For some service organisations, the identification of those sharing the fruits of the organisation's efforts, can be a difficult task.

4 Customer satisfaction

Long term satisfaction of customer needs will be an aim of any Total Quality organisation.

Explanation and description of principle.

In most organisations, the aim is to be in a long term relationship with the customer, albeit this may be distant. Customer satisfaction becomes the nub of the Total Quality organisation.

5 Aim

A Total Quality organisation will have a clearly stated, widely understood and generally accepted *direction / aim.*

Explanation and description of principle.

This direction may be expressed in a variety of ways, including possibly a Mission Statement or publicly stated Policy Document.

6 Communication

A Total Quality organisation will openly and clearly communicate its *principles / beliefs / values / mission statement / policy for quality.*

Explanation and description of principle.

The intention of the organisation, as a whole, must be explicit, providing a business relationship for those outside the organisation, and common understanding and a working framework for those within it.

7 Ethos

Total Quality Management embodies the *values / beliefs / ethos* of the organisation, and thus Total Quality is intrinsic to every activity, decision and action.

Explanation and description of principle.

TQM cannot be an additional activity, an add-on or a subsidiary consideration, in that it is how the organisation is managed; it is the organisation.

Clearly, the ethos, values and beliefs of the organisation, and of those senior personnel, who most influence the organisation's stance, must be compatible with Total Quality.

These senior personnel must promote actions and exhibit values and beliefs, that are in accord with Total Quality, for the organisation to find success from this concept.

8 Values

The highest levels of integrity, honesty, trust and openness are essential ingredients of Total Quality Management.

Explanation and description of principle.

The most significant values, within the Total Quality organisation, are probably expressed within the range of integrity, honesty, trust and openness.

9 Mutual respect and benefit

There is an implicit mutual respect of all stake-holders involved with a Total Quality organisation, that assumes that long term business is intended to be mutually beneficial to all concerned.

Explanation and description of principle.

The future well-being of everyone involved with a Total Quality organisation, requires each individual to maximise their contribution to that future.

Mutual respect, mutual trust and mutual benefit are inextricably entwined and are important factors within the development of any Total Quality organisation.

10 Health and safety

Health, safety and environmental issues have a high priority within a Total Quality organisation since the welfare of all *investors / employees / suppliers/ the community*, as stake-holders in the enterprise, is intrinsic to the future well-being of the organisation.

Explanation and description of principle.

Mutual benefit is expressed through the care taken and respect held for those involved with, and influenced by, the conduct of the organisation.

11 Commitment

Leadership of Total Quality Management stems from the top of the organisation and enlists individual and team commitment throughout.

Explanation and description of principle.

TQM embodies a working partnership.

This is usually achieved by empowerment of those directly involved with an activity, to enable them to be responsible for the Quality outputs of that activity.

TQM will not occur by accident, and there is a considerable responsibility for professional management skills to be consistently used throughout the organisation.

People must be aware of what is expected of them and be equipped to meet these expectations.

12 Participation and ownership

Total Quality offers each individual the opportunity to participate in, and to feel ownership of, his/her activities, and to jointly share a sense of ownership for the success of the entire organisation.

Explanation and description of principle.

Full participation and a sense of ownership can never be successfully demanded; they are given willingly or not at all.

However, the provision of the opportunity for each individual to participate, contribute and develop a sense of ownership, is a key aspect of any successful Total Quality organisation.

Ownership of process allows an individual to make decisions and thus improve on-going control.

Total participation in continuous improvement, will be implicit in most Total Quality organisations.

13 Continuous improvement

Total Quality Management involves continuous and measurable improvement at all levels of the organisation, ranging from company performance to individual employee performance, such that continuous process improvement, forever, becomes an essential ingredient of success.

Explanation and description of principle.

TQM concentrates upon processes and process control.

Continuous improvement of all processes becomes a way of life.

The measurement of such improvement, whether at company, department or individual level, is essential, if the improvements are to be sustained.

Continuous improvement will also take place through innovation, research and development.

14 Performance

Total Quality Management requires <u>*consistent / predictable / accurate / precise*</u> performance to high standards in all areas of the organisation, thus measurement, assessment and auditing are common TQM activities.

Explanation and description of principle.

Many organisations obtain certification to BS 5750, parts 1, 2 or 3, ISO 9000 series, or similar Quality standards en route to Total Quality. Those that don't, nevertheless, adopt very similar methods of documenting procedures, measuring activities and auditing their processes.

Objective assessment and auditing of performance does not contradict other principles of openness and mutual respect, provided it is institutionalised and part of a continuous improvement programme.

15 Resources

An aim of every Total Quality organisation is to better use resources, to achieve greater success, financial and/or otherwise.

Explanation and description of principle.

Every organisation, whether providing a product, service or a mixture of both, is operating to provide added value, to the resources it has available.

The public services' relationship with financial success may be somewhat more complicated, but is, nevertheless, still strong.

16 Investment

Total Quality Management will always require *sufficient / appropriate* investment to ensure that planned activity can occur.

Explanation and description of principle.

TQM requires investment, that is strategically targeted to produce the Total Quality direction identified by the organisation. This investment may well provide a quick return from lower operating costs, improved productivity and increased customer demand.

Clearly the potential improvement must match the investment, and the investment must match the potential added value, to be gained from the improvement.

The term investment will have wider meaning than just a financial input.

Index